'Exploring the nexus of mining, mine closure and housing is an important undertaking because the public, private and civil society sectors are grappling with better arrangements for establishing viable post-mining economies. The book is quite timely since no other book appeared on this topic during the past 30 years addressing this issue in SA. The book is a must read for people working in the fields of urban economics, urban and regional planning, sustainable regional development, housing studies, urban sociology, human geography, etc. and with an interest in housing and mining. Many managers, operating in the mining sector, may find this book useful to inform the policies and plans they design and implement in arranging housing in a mining context and post-mining context. The book is a valuable contribution for the so-called developing world because of the sheer numbers and proportions of populations who remained affected after the extracting industries have closed and left a region.'

Lucius Botes, *Professor in Development Studies and Director Research Development, Faculty of Economic and Management Sciences at the North West University, South Africa*

'Mines come and go, as governments come and go, and people move in and out. Very rarely, these cycles are perfectly attuned, and the expectations for mining as a provider of community prosperity and stability are all too often disappointed. In this impressive synthesis, Lochner Marais takes a hard look at the situation of South African mining towns, and what happens when the mine closes. His conclusions end up far from the commonplaces pervading the literature, arguing expertly on two sides of common polarizations such as camp/town, rental/home ownership, development/shrink, and considering the pros and cons of place attachment.'

Kristof Van Assche, *Professor Planning, Governance & Development, Department of Earth & Atmospheric Science, University of Alberta*

The Social Impacts of Mine Closure in South Africa

This book investigates the relationship between mining, mine closure and housing policy in post-apartheid South Africa, using concepts from new institutional economics and evolutionary governance theory.

Mine closures present a major challenge to the mining industry and governments, with this being particularly noticeable in the Global South. This book argues that the dependencies created by the mining industry and mine housing policies while a mine is operational cause serious societal problems when it closes. To demonstrate this, the book applies the concepts of place attachment, asset-based development and social disruption. Conceptually, the book challenges the view that place attachment and asset-based development are the most appropriate and often the only policy responses in mining areas. In South Africa, the mining industry and the government have created comprehensive housing programmes linked to homeownership to promote place attachment, stability and wealth among mine workers. These programmes do not consider the disruption that mine closure might bring. The book challenges the blind application, during boom periods, of policies which create long-term dependencies that are difficult to manage when a mine closes.

This book will be of interest to students and scholars researching the social impacts of mining and the extractive industries, social geography and sustainable development, as well as policymakers and practitioners working with mine closure or social impact assessments.

Lochner Marais is Professor of Development Studies at the Centre for Development Support and Associate to the Chair on City-Region Economics at the University of the Free State, South Africa. He is also Adjunct Honorary Professor at the Minerals Institute at the University of Queensland in Brisbane, Australia. He is the co-editor of *Coal and Energy in South Africa* (2022), *Mining and Community in the South African Platinum Belt* (2021) and *Community and Mining in South Africa* (Routledge, 2018).

Routledge Studies of the Extractive Industries and Sustainable Development

Our Extractive Age
Expressions of Violence and Resistance
Edited by Judith Shapiro and John-Andrew McNeish

The Impact of Mining Lifecycles in Mongolia and Kyrgyzstan
Political, Social, Environmental and Cultural Contexts
Edited by Troy Sternberg, Kemel Toktomushev and Byambabaatar Ichinkhorloo

Oil and National Identity in the Kurdistan Region of Iraq
Conflicts at the Frontier of Petro-Capitalism
Alessandro Tinti

The Anthropology of Resource Extraction
Edited by Lorenzo D'Angelo and Robert Jan Pijpers

Andean States and the Resource Curse
Institutional Change in Extractive Economies
Edited by Gerardo Damonte and Bettina Schorr

Stakeholders, Sustainable Development Policies and the Coal Mining Industry
Perspectives from Europe and the Commonwealth of Independent States
Izabela Jonek-Kowalska, Radosław Wolniak, Oksana A. Marinina and Tatyana V. Ponomarenko

The Social Impacts of Mine Closure in South Africa
Housing Policy and Place Attachment
Lochner Marais

For more information about this series, please visit: www.routledge.com/Routledge-Studies-of-the-Extractive-Industries-and-Sustainable-Development/book-series/REISD.

The Social Impacts of Mine Closure in South Africa

Housing Policy and Place Attachment

Lochner Marais

Routledge
Taylor & Francis Group
LONDON AND NEW YORK

earthscan
from Routledge

First published 2023
by Routledge
4 Park Square, Milton Park, Abingdon, Oxon OX14 4RN

and by Routledge
605 Third Avenue, New York, NY 10158

Routledge is an imprint of the Taylor & Francis Group, an informa business

British Library Cataloguing-in-Publication Data
A catalogue record for this book is available from the British Library

Library of Congress Cataloging-in-Publication Data
Names: Marais, Lochner, author.
Title: The social impacts of mine closure in South Africa : housing policy and place attachment / Lochner Marais.
Description: New York : Routledge, 2022. |
Includes bibliographical references and index.
Identifiers: LCCN 2022009951 (print) | LCCN 2022009952 (ebook) | ISBN 9781032200552 (hardback) | ISBN 9781032325743 (paperback) | ISBN 9781003262015 (ebook)
Subjects: LCSH: Housing—South Africa. | Housing policy—South Africa. | Mine closures—Social aspects—South Africa.
Classification: LCC HD7374.4.A3 M37 2022 (print) |
LCC HD7374.4.A3 (ebook) |
DDC 363.510968—dc23/eng/20220225
LC record available at https://lccn.loc.gov/2022009951
LC ebook record available at https://lccn.loc.gov/2022009952

ISBN: 978-1-032-20055-2 (hbk)
ISBN: 978-1-032-32574-3 (pbk)
ISBN: 978-1-003-26201-5 (ebk)

DOI: 10.4324/9781003262015

Typeset in Goudy
by codeMantra

Contents

Figures

Tables

Annexures

Foreword

Research about life on the fringes of society is incredibly important. The fringe is a tenuous place to exist. Here, people can hit just one snag, and life can unravel. Losing a job, a home, a property, one's health or access to social support can pull lives and livelihoods apart to devastating effect. Some people exist on a different plane, where the fabric of society keeps them far from the fringe. Some don't see the fringe from their privileged position. Others refuse to think about what it is, how it is formed, how it works and whether they have a role in maintaining or changing it.

At certain times, and especially in moments of crisis, people with power and privilege may recognise the perils of a social and political system that keeps some people safe and makes others vulnerable. Unfortunately, even when the perils are recognised, attempts to repair the social fabric can become entangled with threads from the past. And so, the problems of old are remade in different forms.

The Social Impacts of Mine Closure in South Africa: Housing Policy and Place Attachment is a book for people seeking to understand entrenched systems of inequality and the fabric of South Africa's mining towns. These towns are integral to South Africa's complex history. They are formed – quite literally – on the edge of the industry, growing and expanding in and around mining footprints. These populations service, support and complicate the industry in ways that often go unrecognised.

Set against the backdrop of the transition to post-apartheid South Africa, the book spends little time describing who exists on the mining fringe. This is well established. South Africa's poor are overwhelmingly black, landless and multidimensionally disadvantaged, that is, not poor by any single measure, but on almost every conceivable measure. Instead, the book works to characterise mining towns and their failings, not only by analysing who is disadvantaged, but also by questioning how and why poverty and deprivation persists, despite the country's rich resource endowment.

I first met Professor Lochner Marais in 2019. My colleague, Professor John Owen, and I had wanted to visit the Centre for Development Support at the University of the Free State for some time. Most of our applied research in mining had been conducted in Australia, the Asia-Pacific and the Americas, and we were eager to learn about other contexts. The Centre's research on human development was renowned, and Lochner's work on mining directly aligned with our interests.

On arrival in Bloemfontein, we visited the city's lookout. As we surveyed the landscape, Lochner narrated the history of place through waves of conflict, colonialism and settlement. Corridors of rural migration and present-day highways, railways and transport routes were pointed out. Various spatial boundaries, where old racial segregations had been retained and new social divisions had formed, were explained. State policies had addressed some of apartheid's legacies by connecting the sprawling informal settlements to essential services, but those same policies had exacerbated other issues, such as urban poverty and crime. Lochner made his knowledge easily accessible to us and, as we witnessed during our visit, to many others.

This generosity continued throughout our stay, especially on our guided tour through nearby resource regions, including the book's case study towns of Tsantsabane, Emalahleni and Rustenburg. We visited informal settlements, abandoned mines and areas cleared for resettlement. We stopped at the town of Virginia, the site of a historical tailings dam disaster that flooded the suburb of Merriespruit, causing multiple fatalities. It was during this trip that we discussed many of the themes that appear in this book and debated the conditions under which mining can be the snag that pulls the whole social fabric apart. Our exchanges motivated several subsequent collaborations with Lochner and his colleagues, including a joint paper, led by John, on 'localising the resource curse'.

The Social Impacts of Mine Closure in South Africa is a forensic work that applies a diverse range of theories, concepts and methods. The book abandons commonly held assumptions that limit our ability to characterise the problems and possibilities of mining towns. For instance, the concept of 'disruption' is used to analyse mine closure, challenging the assumption that closure signals the end of mining's disruptive impacts. Positive notions of 'place attachment' are rejected. In contexts where mine dependency is high and economic decline is inevitable, home ownership and associated debt can exacerbate household poverty. The positive potential of informal housing, widely assumed to be negative, is carefully considered. The book posits that well-managed impermanent housing might avoid the 'locked-in' effect of home ownership. Even the modern mantra of 'economic diversification' becomes a false pursuit – a distraction invented by corporate and state actors to avoid provisioning for the inevitable economic decline that accompanies mine closure.

Despite its applied work in the mining sector, the Centre for Development Support is itself relatively marginal in the world of mining. This is not to say that the Centre is doing marginal work; far from it. Lochner's book is a testament to the independent, applied scholarship that it pursues. Research about mining's adverse impacts is rarely supported by industry or governments. It is difficult to hold steady a course of inquiry when the weight of pro-mining opinion pulls entirely in the other direction. Research topics will fall in and out of favour, yet some remain constant in their importance. This book is offered in that vein: it is not written to appeal to powerful actors, but to challenge them.

For those unfamiliar with the work of the Centre, I encourage you to find out more. You may struggle, however, to familiarise yourself with Lochner's body of work. He is prolific, and his writings are dispersed amongst numerous topics and collaborative endeavours. Much of his intellectual energy is directed towards enabling students and early career researchers at the Centre and joint works with his peers. It is thus rare to have direct access to Lochner's thinking. In this book, he has built on the work of others by using available data to craft his own piercing analysis about the dilemmas of mining and development.

If companies and governments are serious about mining's contribution to development, they could use their vast resources to digest alternative arguments such as those in this book. Instead, they invest in trying to convince communities that, in the end, mining will leave them better off. The book confirms how wide the knowledge gap has become. If we are to have an impact as scholars, we would do well to focus as Lochner does on characterising problems and bridging the gap. There is too little success to be found on the fringes of mining, and we must carve out intellectual space to interrogate why this continues to be the case.

Professor Deanna Kemp
Director, Centre for Social Responsibility in Mining
Sustainable Minerals Institute
The University of Queensland

Preface

A memorable field trip I took as a student at the University of the Free State 30 years ago sparked the interest that led to this book. It was an outing for the honours class to the Free State Goldfields in August 1990, arranged by our enterprising lecturer in political geography, Skip Krige. The apartheid government had unbanned the African National Congress (ANC) a few months previously, and mass land invasions were happening across the Goldfields. Some farmers had taken up 'shack farming' to provide housing for the families who were now able to visit their men working on the mines. We visited Beatrix Mine, where Bernard Swanepoel, at the time the youngest mine manager in the history of South Africa, spoke to us about the future of mining. I am sure that what he had to say impressed me, but I was even more impressed by the splendid evening we had at the Beatrix Mine's guest house in Virginia. We were regaled with more food and drink than even ten hungry students could do justice to. This was the highlight of the trip for us. But more importantly, I was left with a lasting interest in housing for South Africa's miners.

Although I tried many times to return to Free State Goldfields, I did not get the chance until 2010 when the Industrial Development Corporation contracted me to investigate the decline of mining and its effects on the local economy. Twenty years on, things had changed. None of the mines had guesthouses anymore. No more five-star treatment. I found a place run down, people struggling to survive and nobody outside of the area willing to listen to stories about the economic hardship of mine closure. I heard many big ideas but saw very little progress. Political manoeuvring overshadowed economic recovery. However, I still found friends who were willing to talk to me. Karel van der Walt and Fanie Niewoudt remain people with open doors and hearts. They lived through the pain and saw their plans disappear. Several postgraduate students have kept me interested in the Free State Goldfields. Kentse Sesele in particular stands out.

Following the research for the IDC, it took me a few years to write up stories of decline and delve into the area's history. Reading and writing about this history taught me one lesson: the effects of mine closure originate from the way mining booms are managed. Early decisions become set in stone and we are stuck with the consequences, just the way the QWERTY keyboard, developed in the late 1800s, determines how I type today. The 1950s garden city plan, with its wide roads, open spaces and low densities, is probably the most significant cost driver today in the Free State Goldfields. In the 1960s, planners hailed this planning as modern and creative. Nobody thought of the consequences of being stuck with it 50 years later. At the heart of this book lies my concern about the rigidities and dependencies that people create during mining and the damage this can do over the years. I saw this for the first time in the Free State Goldfields.

On a warm Friday afternoon in December 2014, I had a beer or two with Philippe Burger and JP Geldenhuys (a much more moderate intake, I might add, than in August 1990 in Virginia). We wanted to discuss a new research plan. The lack of data about mineworkers was a central concern in this discussion. It gave birth to the mining town's project. This project has delivered three edited collections (on Postmasburg, now Tsantsabane, Emalahleni and Rustenburg) and numerous articles. The three books enabled us to interact with international scholars like Etienne Nel (New Zealand), Fiona Haslam McKenzie, John Owen and Deanna Kemp (Australia), Franklin Obeng-Odoom (Ghana and Finland) and Kristof van Assche and Leith Deacon (Canada). The National Research Foundation provided small grants to get some of these international scholars to South Africa and pay their expenses. A particular word of thanks to Deanna Kemp, who paid most of her and John Owen's costs on their visit. My sincere gratitude to colleagues who are probably tired of listening to my arguments but keep on indulging me. Colleagues who answer my emails because I nag. Colleagues who send me a smiley face when I am on the verge of exploding. Your kindness does not go unnoticed. Thanks to those who did key informant interviews: Deidre van Rooyen, Phia van der Watt, Stuart Denoon-Stevens and Verna Nel.

Writing a book like this would not have been possible without four people. Molefi Lenka managed the three large household surveys. I do not know how he does this. I am too scared to ask. I know it takes commitment and endurance. Without Molefi, the book would not have had the empirical richness I wanted to bring to it. Of course, no household survey is perfect, but Molefi's efforts are very close to it. Second, Jan Cloete manages the data when it arrives through clouds and other virtual processes. Again, I have no idea how he does this. He also answers my many emails without complaining and often tells me politely that I am wrong and do not understand data. I sometimes listen to him. Then there is my language editor, Di Kilpert, someone I try to always listen to but usually fail. The amount of detail she picks up is mind-boggling. She did it again for this manuscript. I will always be indebted to her. I am also much indebted to my dean, Hendri Kroukamp, who approved a sabbatical in 2021 without asking too many questions.

And then there are the people at home who listen to my crazy ideas. They often nod to say they understand, even when I sense this is due to politeness. Thanks again.

Lochner Marais
University of the Free State, Bloemfontein

1 Setting the scene

1.1 Background

The QWERTY keyboard is a classic example of how we come to depend on an outdated idea. Remington, makers of the first typewriters, spent years looking for the best way to distribute the alphabet to stop the keys jamming. Eventually they settled for the QWERTY arrangement in 1873, familiar on our keyboards today. But computers do not have mechanical keys that can get tangled. The letters could be distributed many other ways to improve typing speed. But we have become so used to the QWERTY arrangement that nobody dares change it.

Mining towns have become similarly dependent on old ideas that are in effect QWERTYs. For many towns, mining is their only viable economic sector. The community comes to depend on the mine and cannot imagine it going bust. Mine decline and closure are not high on their agenda: they prefer to trust their QWERTY. My analogy is not perfect, of course. The QWERTY keyboard works. It causes no ill effects. But the old ideas that mining towns depend on can be dangerous.[1]

Mining dates back thousands of years but mass extraction of minerals only about 200. Company towns sprang up as the multitudes responded to the lure of the mines. The companies took charge and exercised control without regard to democratic and public institutions. Today, the mines are moving away from the company town model but the economic dependence remains. When a mine closes, the community find themselves living in a town whose economy is going downhill. Accustomed to mine employment, few have the capacity to find other livelihoods, and the municipality is not geared to taking over where the mining company left off.

This book explores the implications of dependency in failing mining towns in South Africa.

1.1.1 From company towns to open towns

A 'company town' is a settlement built and operated by a single business enterprise (Garner, 1992). The term was coined in the late 1800s and initially included

DOI: 10.4324/9781003262015-1

workers' camps. Company towns are common in the mining industry: between 1830 and 1930, this was the way most mining companies housed their workforce and many of these towns still exist. These towns ensured adequate labour would be available in the mining area as they provided housing and services at low cost. Many books and papers have been published on company towns (Porteous, 1970; Garner, 1992; Green, 2011). Two themes are prominent: the cultures and subcultures associated with mining (Schumann, 2020) and the dominant role of the company. Garner (1992, p. 4) notes that 'virtually everything associated with the settlement, including the houses, store, school, and even the chapel, was subordinate to the business enterprise'.

After the 1930s, there was a steady shift away from company towns, and by the 1970s, the pressure to reconsider this model had mounted. Mines stopped creating new company towns, normalised the existing ones and worked with democratically elected councils to manage them. But in Africa, many company towns remain (Gough et al., 2018). The change in approach was for four reasons. Some countries switched to the fly-in-fly-out model, reducing the need for housing at the place of mining and to some extent making mining localities less dependent on mining booms and busts (Haslam McKenzie, 2016, 2020). Tax policies changed and taxes were imposed on the fringe benefits of cheap company-provided housing (Marais et al., 2018b). Governments and civil society began to object to private provision of public services (Obeng-Odoom, 2009). And globalisation and the pressure from international shareholders obliged the mining companies to find ways to minimise the cost of peripheral activities like housing (Marais et al., 2018a). By the mid-2000s, the mining industry was actively promoting open towns and normalising existing company towns (IIED, 2002). The dominant role of the mining companies gave way to collaborative planning. Between them, the mining companies and democratic local governments had to ensure appropriate local governance for mining towns.

Very few mining companies still create company towns. Where fly-in-fly-out arrangements are not possible, collaborative planning has become the preferred approach. Collaborative planning requires appropriate local capacity to ensure that mining benefits both the mineworkers and the community. But the plight of mining towns newly responsible for their own planning, albeit in collaboration with mining companies, is not always well understood. Having ensured more local and democratic processes, open and normalised towns now find themselves responsible for managing the risks of mine closure. Collaborative planning transfers this responsibility to the local communities and individuals. Canada has a long history of local governments taking responsibility for mining booms and busts (Deacon et al., 2018; Van Assche et al., 2020, 2021). Dealing with decline is not easy. The fear of becoming a ghost town is a theme prominent in the US literature (Dallas, 1988; Smith, 2011).

The move away from company towns to normalised or open towns is a global phenomenon. Governments, civil society and mining companies have all supported this move. But this shift in policy has not changed the dependence on mining: it has merely made mining communities and individuals more directly responsible for dealing with the consequences of mine closure. In effect, it has reinforced their dependence on mining.

1.1.2 Mine housing in South Africa

South Africa has had its fair share of company towns (Mabin, 1993) and normalisation (Van der Watt & Marais, 2019). Originally, most mineworkers were housed in single-sex hostels at the mines. A system of institutionalised migrant labour applied to the black workforce, as part of the broader racial policies of apartheid (Crush, 1989, 1994). The first single-sex hostels were built at the Kimberley diamond mines in the late 1800s, designed primarily to ensure control and minimise theft. In contrast, white mineworkers had a choice of finding their own housing or living in company housing, which was readily available in both open and company towns (Mabin, 1993). When homeownership for black workers became possible by the mid-1980s, research focused on the likelihood of black mineworkers taking up homeownership and obtaining mortgages (Laburn-Peart, 1990; Crush, 1992).

By the early 1990s, the housing options for black mineworkers had become more diverse. Unions could negotiate a living-out allowance for them. The allowance consisted of an amount added to their salary if they chose not to live in a hostel. With the scrapping of the Group Areas Act in 1991, black mineworkers could also access company housing that had originally been only for white workers. As living-out allowances became more common, black mineworkers were able to find housing for themselves. However, progress was slow, and disrupted by mine closure in the gold industry (Marais, 2013a). The banks redlined some of the gold mining areas. For example, by the mid-1990s, they were not providing mortgages to Thabong, the former black township near Welkom in the Free State Goldfields (Tomlinson, 1997).

The post-apartheid government objected to the institutionalised migrant labour and single-sex hostels. The mining companies had to dismantle their hostels and replace them with homeownership and housing in open towns. Post-apartheid policy promoted the opposite of the previous regime's policy, with obvious moral justification. However, there are several concerns about the new policy. A model dominated by homeownership assumes that migrant labour will not continue. Homeownership assumes long-term economic stability, which is often doubtful in the mining sector. Rental housing is an option little used in mining towns, despite ample rental housing being available. The single-sex hostel system was inhumane, but a point in its favour that we seldom hear about is that it was a form of mass rental housing (with the company paying the rentals).

In summary, post-apartheid housing policy for mining areas tried to redress the long history of racial discrimination in housing policy and the consequences of institutionalised migrant labour. But migrant labour is one of those dependencies that remain with the mining industry. The emphasis on homeownership, open towns and normalisation does not consider what will happen to homeowners if a mine closes.

1.1.3 Mine closure

Like death and taxes, mine closure is certain. But predicting which mines will close and when is not easy. The predictions would have to be based on the quantities of resources remaining and market volatility. In 2002, the World Bank estimated that 25 large mines in the developing world would close in the next decade. Nobody verified these predictions and further predictions have not been forthcoming. Much information is available on resource reserves but little on mine closure. In particular, the body of research on the social effects of mine closure remains small compared with the extensive work that has been done on the environmental and legal aspects.

I was able to find only three books on mine closure, one single-authored (Chaloping-March, 2017) and two edited collections (Neil et al., 1992; Keeling & Sandlos, 2015). But there has been an upsurge in a variety of other literature (Strongman, 1992; Warhurst et al., 1999; Jackson, 2002; Laurence, 2002; Haney & Shkaratan, 2003; McGuire, 2003; Rao & Pathak, 2009; Digby, 2012; Manjunath et al., 2016; Bainton & Holcombe, 2018; World Bank, 2018; Vivoda et al., 2019). The research profile in South Africa is similar. Macmillan (2012, p. 547) found it strange that 'such dramatic changes in the nature of South Africa's gold-mining industry, and patterns of employment in an area like the Free State, have attracted so little academic, or journalistic, attention'. Since Macmillan's observation, more studies on mine closure have been published in this country (Marais, 2013a, 2013b; Marais & Cloete, 2013; Siyongwana, 2018; Marais & De Lange, 2021; Sesele et al., 2021). Watson and Olalade (2019) and Crous et al. (2021) have investigated the possibility and likely scale of mine closures in South Africa, with the former highlighting the government's slow processes in enabling mine closure. A recent edited collection on the likelihood of mine closure in Emalahleni has broadened the debate (Marais et al., 2022a).

1.2 Mine housing policy in South Africa

Three features of mine housing under apartheid were migrant labour, single-sex hostels for black workers and company houses for white workers. Post-apartheid policy, supported by the unions, aimed to change all this. The mines had to demolish the single-sex hostels, dismantle institutionalised migrant labour and create parity between black and white mineworkers. The three main policy directions were homeownership, open towns (or normalisation of existing company towns) and the elimination of racial discrimination in housing policies and practices. An 'open town' is one that hosts both mineworkers and non-mineworkers. 'Normalisation' means opening a company town to other residents. In practice, it means the mining company hands over the management of the town to public governance.

In some cases, it means closing a company town and requiring residents to move to an open town (Van der Watt & Marais, 2019). Homeownership became available for black mineworkers from the mid-1980s and the mines started to sell their housing stock to their mineworkers and on the open market. Ownership became the dominant form of housing tenure advocated in the mining industry for three reasons: to counter the migrant labour system, as an alternative to the single-sex hostel, and to contribute to mining areas' local economic development. Making homeownership available would mean that migration trends would subside. The International Council of Mining and Metals (ICMM) propagated ownership and open towns from the 1980s and recommended collaborative planning to counter the dominant approach by mining companies (IIED, 2002). In South Africa, collaborative planning suited the democratic dispensation's promotion of developmental local government. Mining policies and guidelines emphasise integrating mining and non-mining households and using mining as a tool to foster local economic development. Living-out allowances would create parity between black and white workers. They would allow thousands of mineworkers to make their own housing choices rather than have decisions made by for them by the mining companies or the government.

Beyond the changes discussed above, post-apartheid housing policy entrenched two ideas: place attachment and social stability based on assets.[2] By institutionalising migrant labour, apartheid policy for mine housing prevented people from becoming emotionally attached to a town. Homeownership, a house of one's own choice and open towns where mining becomes a local economic development tool were attempts to increase place attachment and social stability through asset creation.

This seems a logical alternative to apartheid policy. It is generally a good thing to foster place attachment and social stability. But the new policies have not been assessed critically in post-apartheid research. Several criticisms can be made. It has not been sufficiently taken into account that homeownership is a good thing only where there is long-term economic prosperity. It is difficult to guarantee long-term economic prosperity in a town dependent on mining. The focus on homeownership distracts people from other tenure options. The success of open towns and normalisation depends on several factors. It assumes adequate local capacity to deal with the impact of mining and possible mine closure. Collaborative planning can be difficult, given the unequal power relationships between the mines and the municipalities (Van der Watt & Marais, 2021). Local economic development assisted by the mines is a good thing, but little thought has been given to how municipalities can benefit from mining. Most seriously, the considerable risks associated with mine closure do not receive much attention when government, mines and unions consider the new policies. The closing of a mine can be socially and economically disruptive when mining has created long-term dependencies for mining communities and households. Insufficient thought has been given to planning for this disruption or finding ways to avoid dependencies developing when mining is growing.

This book challenges the simple assumptions in post-apartheid housing policy by asking some pertinent questions, such as: How much place attachment and stability have the new policies created in towns where mines are very likely to close? What dependencies have contributed to post-apartheid housing and settlement policies in mining areas and have these policies created new dependencies? What are the implications of mining and mine closure for housing policy and settlement development near mines?

1.3 The scope of this book

This book investigates the long-term implications of South Africa's policies for mine housing, especially in towns threatened by mine closure. The presence of a mine gives a town a false sense of security, leading governors and planners to assume that the prevailing economic prosperity will continue. I refer to this failure to imagine a different future as 'dependency', a term I borrow from new institutional economics (North, 1990, 2005) and evolutionary governance theory (Van Assche et al., 2014, 2016).

I examine a wide range of issues: apartheid and post-apartheid policies and the dependencies associated with current and future policies; the attributes and inequalities of housing in mining areas; the growth of informal settlements in mining areas; housing and migration responses from mining companies and households in mining towns; the role of place attachment and asset-generation in housing policies for mining communities; the concept of mine closure and its link to social disruption; and the implications of mine closure for housing policy. I create a conceptual framework for thinking about housing, mining and closure which will contribute to dependency theories. Finally, I present a set of policy proposals for mine housing in South Africa.

1.4 Conceptual background

1.4.1 *Theory*

My discussion draws on two strands of theory: new institutional economics and evolutionary governance theory.

New institutional economics developed as a response to classical and neoclassical economics, which emphasised capital and efficiencies. Coase's original works, *The Nature of the Firm* (1937) and the *Problem of Social Costs* (1960), laid the foundation. Williamson used the term 'new institutional economics' for the first time in 1975 (Williamson, 1975). New institutional economics focuses mainly on 'the rules of the game' and how the rules change, leading to institutional change (North, 1990). North (2005) says change is often slow or path-dependent. This theory's concept of dependence, and particularly path dependence, has been useful for my analysis of policy and governance.

Evolutionary governance theory draws on other theories, such as biological evolutionary theory, social systems theory, actor-network theory, post-structuralism and new institutional economics (Van Assche et al., 2014). A contribution from the field of biology is the concept of *autopoiesis* (self-creation) (Maturana & Varela, 1972; King, 1993; Von Krogh & Roos, 1995). Transferred from the natural to the social sciences, this concept has proven useful for studying social systems, such as informal settlements in mining areas (Marais et al., 2022b). Evolutionary governance theory also draws on Foucault's ideas about discourse analysis and power (Foucault, 1982), seeing the evolution of policies as an effect of power relations. Evolutionary governance theory resembles more recent theories on network governance and has become a dominant approach to governance studies (Klijn, 2008). Many of its ideas are relevant to the topics in this book, such as its view of governance as changing or evolving, its emphasis on non-linearity between cause and effect or between two elements in a system and its focus on the co-evolution of governance (policy seen as resulting from the interaction between the actors in the decision-making process). Evolutionary governance theory occupies the middle ground between theories of social engineering and theories of free market economies. It provides a direct link between governance and politics.

The analysis in this book owes much to the three dependencies distinguished by evolutionary governance theory and new institutional economics: path dependency, goal dependency and interdependency.

The concept of *path dependency* emphasises the fact that history matters. The term is connected with the term 'historical institutionalism', first used by political scientists in the mid-1980s (Krasner, 1984). New institutional economics made path dependency a central concept to explain economic and technological change (North, 1990). Van Assche et al. (2016) use it as one of the building blocks of evolutionary governance theory. It is important to understand that path dependency does not predict the future but shows that recent decisions have historical roots. The focus in path dependency is on the evolving nature of decision-making. Policy decisions can become governance paths and these paths depend on specific cultural and historical contexts (Ubels et al., 2019). Previous policies create preconditions for the reproduction of existing governance paths. In this way, past decisions transfer elements of historical decision-making into the future. The focus on history helps to explain and compare alternatives. The concept of *autopoiesis* helps to explain how policy decisions spring from history reproducing and maintaining itself. A new path usually begins at a critical juncture. It could lead to many different destinations. The result might not be linear or aligned with the intent of the original decision. North (2005) coined the term 'lock-in' to describe slow change (or inability to change) and people's inability to escape history and rigidities. He distinguishes between institutional changes (rule changes) and technological changes and points out that institutional change is usually more difficult to achieve than technological change.

Goal dependency refers to the negative implications of a shared vision of the future that could affect present decision-making (Van Assche et al., 2014). Few people would argue that a shared vision is a bad thing, but evolutionary governance theory emphasises the unintended consequences. A shared vision can lead to inability to change and inflexibility in policy and planning. The world's alignment to the UN Sustainable Development Goals is an example of goal dependency on a global scale. Many countries might implement the goals differently from the original intention because of self-interest or inflexibility, or the domination of power groups.

Various theories, such as social systems theory (Luhman, 1995) and new institutional economics (North, 2005; Greiff, 2006), recognise the reality of *interdependency*. Evolutionary governance theory sees interdependency as stemming from the historical relationships between actors and institutions. These relationships can create rigidities because of path dependency or they can create resources, expertise and flexibility. Central elements of interdependency include an interrelationship between trust, co-operation, power and conflict (Van Dijk & Vermunt, 2000; De Vries & Nemec, 2013).

1.4.2 The policy debate

Mining has generated extensive policy debates in South Africa, particularly about mine ownership and how to achieve a more equitable shareholding structure while remaining internationally competitive. These debates are necessary but they are not the focus of this book. There has been little debate about the link between mine housing and South African housing policy. Post-apartheid policy demanded the dismantling of single-sex hostels and no one would call this move inappropriate.

However, two interrelated aspects of post-apartheid mining policy have gone unchallenged: the drive for homeownership and open towns, which assumes collaborative planning, and the drive to link mining to local development. It is in this specific policy space that I would like to contribute with this book. Permanency through homeownership ceases to be a good idea when mine decline sets in. It is the path dependency that new growth brings to a mining town that later makes it difficult to change direction. And it is the goal dependency created by the policy for redressing historical ills that lies at the heart of the problem.

In this book, I attempt to answer some questions about the current policy, such as: Why has homeownership become such a dominant policy approach in mining areas? What are the alternative tenure options? How should mining companies and government think about mine housing when a town is threatened by potential mine closure?

Apartheid policy did not consider it important for mineworkers to develop place attachment; rather, the reverse: place aversion would be the more likely effect of living in the hostels. Similarly, workers were not encouraged, or even allowed, to accumulate housing assets in the mining towns. Post-apartheid policy has swung to the opposite pole. The question of whether this is good policy in the context of mine decline and closure is central to this book.

Path and goal dependencies continue into the post-apartheid era. Migrant labour, a path that began under colonial and then apartheid rule, has become a dependency today. The policies for homeownership, mining linked to development, the dismantling of the hostels and the normalisation of towns all create place attachment and housing assets. These policies are an example of a new goal dependency that will eventually create new path dependencies because they do not consider the disruption that mine closure will cause. The new path dependencies could lock households into locations with very little long-term economic viability. Under apartheid, government and mining companies were the primary decision-makers. The government's policies were centrally controlled. Today, the relationships have changed. Unions were legal in the latter days of apartheid but gained power only after 1994. The decision-making power has changed. Under apartheid the government decided where people were allowed to live; under the post-apartheid dispensation, many government regulations have been relaxed and in most cases mineworkers can now make their own housing decisions.

Figure 1.1 shows a framework for the book based on the discussion above. The diagram integrates two theoretical constructs (place attachment, and housing assets for creating social stability), the three dependencies borrowed from evolutionary governance theory and new institutional economics, and housing policy for mining towns in South Africa.

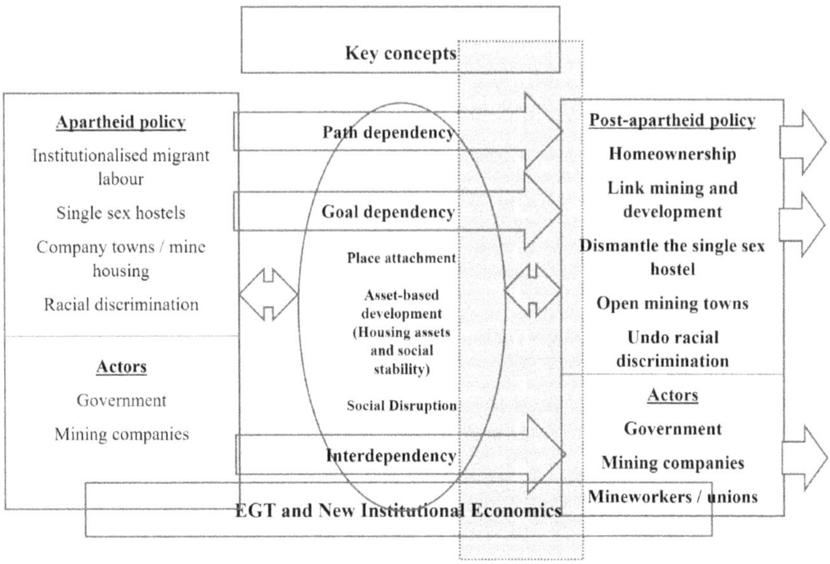

Figure 1.1 Conceptual framework for the book.

1.5 The research that went into this book

The research team conducted case studies to compare three mining towns in South Africa: Tsantsabane, Emalahleni and Rustenburg. The book builds on the three edited collections that resulted from these studies (Marais et al., 2018a, 2021, 2022a). The consequences of mine closure were central to all three studies. Some chapters in these books investigate mine housing and housing policy. In this book, I expand the debates and further investigate the link between mine housing policy and mine decline.

In addition to the case studies, I use four main sources of secondary data. I make extensive use of the household surveys that were conducted by the University of the Free State in the main urban areas of the case study cities: in 2015 for Tsantsabane, 2017 for Emalahleni and 2018 for Rustenburg. Some of the papers that originated from these surveys used inferential statistics, but although I refer to some of this work, I use only the descriptive statistics. I use the 80 key informant interviews that various team members conducted in the three case study areas, which focused on mine-community relationships but also made some reference to mine closure and housing policy. I use Statistics South Africa's census data and community survey data extensively. And I use Global Insight, a valuable source of data to understand economic processes.

1.6 The case study areas

Most mining town research focuses on a single sector of the industry. This book compares towns involved in three different sectors: iron ore (Tsantsabane), coal (Emalahleni) and platinum (Rustenburg). The section on Emalahleni is more detailed, because that study included the important issues of energy from coal, renewable energy sources and what is known as a just transition. Figure 1.2 shows the locations of these towns and Table 1.1 gives their details.

1.6.1 Tsantsabane

The subject of this case study was Postmasburg, a small town in a remote part of the Northern Cape Province, which together with several small surrounding villages makes up the Tsantsabane Local Municipality. In 1892, Postmasburg was established as a regional service town for the surrounding farming communities. Khoi- and Tswana-speaking people mined historically in the area, and the first commercial manganese mining in the area started in 1929. During the 1930s, manganese was the main resource mined in Postmasburg, but these operations remained low-key. The development of the Beeshoek iron ore mine about 20 km outside Postmasburg in 1960 was a step forward for the town. Since 1999, the increasing demand from China for iron ore has stimulated new growth in Postmasburg. Needing to expand, the Beeshoek closed its company town and resettled its workers in Postmasburg through a housing programme providing homeownership. In 2011, Kumba Iron Ore

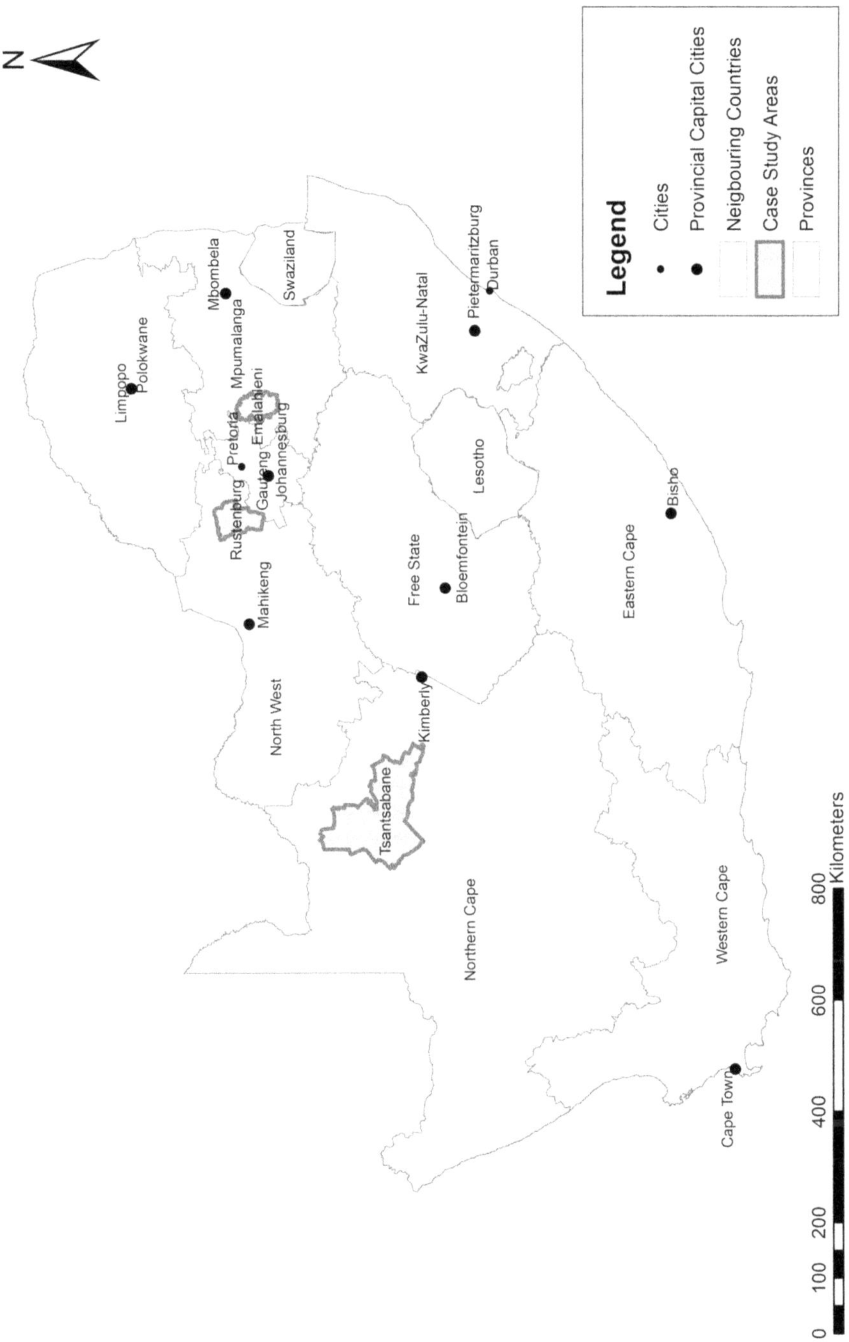

Figure 1.2 Location of case study areas.

Table 1.1 A comparison of the three case study areas

Attributes	*Tsantsabane (2015)*	*Emalahleni (2017)*	*Rustenburg (2018)*
Name of municipality	Tsantsabane	Emalahleni	Rustenburg
Name of main urban area	Postmasburg	Emalahleni (previously Witbank)	Rustenburg
Main resource mined	Iron ore (and some manganese)	Coal	Platinum and related minerals
Type of mining	Surface (opencast)	Surface (opencast) and underground	Underground
Date of first mining	1930s (manganese), 1960s (iron ore)	Late 1800s	1930s
Latest boom	2009–2019	2000–2019	2000–2008
The driver of the boom	Export market for iron ore to China	Global price of coal	International policies on vehicle emissions have raised demand for platinum used in catalytic converters
Long-term risk associated with the resource	Decline likely to occur only in the longer term. Decline in economic growth in China	Decline in global demand for coal and South Africa's shift away from coal to renewables	Move towards electric cars likely to decrease demand for platinum
Population 1996	26,567	236,040	311,787
Population 2001	27,082	276,413	387,096
Population 2011	35,094	395,466	549,595
Population 2016 (estimate)	39,345	455,228	626,522
Population growth per annum, 1996–2001	0.48	3.21	4.42
Population growth per annum, 2001–2011	2.63	3.65	3.57
Population growth per annum, 2011–2016	2.31	2.86	2.66
Population growth per annum, 1996–2016	1.98	3.34	3.55
Mining GVA, 1996 (2010 price × 1000)	1,560,395	14,262,701	18,988,581
Mining GVA, 2001 (2010 price × 1000)	1,651,066	16,518,725	19,252,967
Mining GVA, 2007 (2010 price × 1000)	1,863,056	20,267,254	26,484,359
Mining GVA, 2013 (2010 price × 1000)	2,320,454	17,445,278	25,760,243
Mining GVA, 2019 (2010 price × 1000)	2,717,990	17,154,377	25,552,123
Mining growth, 1996–2001	1.13	2.98	0.27

Mining growth, 2001–2007	2.03	3.46	5.46
Mining growth, 2007–2013	3.72	−2.46	−0.46
Mining growth, 2013–2019	2.67	−0.28	−0.14
Mining growth, 1996–2019	2.44	0.81	1.30
Mining's share of the economy in 1996	17.3	27.3	42.0
Mining's share of the economy in 2001	21.3	24.1	36.2
Mining's share of the economy in 2007	32.3	22.1	48.6
Mining's share of the economy in 2013	50.4	27.4	49.1
Mining's share of the economy in 2019	52.9	28.5	39.5
Gross domestic product (GDP) growth, 1996–2001	1.10	3.89	3.68
GDP growth, 2001–2007	3.18	3.04	4.95
GDP growth, 2007–2013	5.35	2.47	0.58
GDP growth 2013–2019	4.23	2.46	0.73
GDP growth, 1996–2019	3.55	2.66	2.42
People working in mining, 1996	1,157	21,151	46,266
People working in mining, 2001	1,500	22,665	47,789
People working in mining, 2007	2,752	24,865	85,643
People working in mining, 2013	5,861	35,716	89,682
People working in mining, 2019	7,893	40,405	75,404
Were there single-sex hostels?	None in Postmasburg. Beeshoek mine had a hostel for black workers, dismantled in the mid-1980s	Yes. Most have been dismantled	Yes. Most have been dismantled. Some still used for contract workers
Were there company towns?	Yes (Beeshoek)	Yes (various company towns next to the mining sites)	No
Did normalisation take place?	Beeshoek was closed in 2014 when the mine wanted to mine the land on which Beeshoek was located	Yes, most company towns were closed in the 1990s and a few were normalised	No, Rustenburg was originally an open town. But some mine-owned houses are available

(Continued)

Attributes	*Tsantsabane (2015)*	*Emalahleni (2017)*	*Rustenburg (2018)*
Development of an open town	The mine facilitated homeownership for Beeshoek residents in Postmasburg	Since closure of the company towns in the 1990s, Emalahleni has developed as an open town. Residents from the company towns had to relocate to Emalahleni town and find their own housing. Relocation was aided by incentives from mines and Eskom	Rustenburg developed mainly as an open town with some company housing being part of the open town
Availability of rental housing owned by the mines	Kumba has built many houses since the opening of the Kolomela mine, which they rent to their employees. Mineworkers can also find their own housing	None	Some mine rental housing still exists, but mines have sold most on the open market
Presence of migrant workers	Very small percentage	A small percentage	Yes
Number of informal houses, 1996	784	12,901	28,637
Number of informal houses, 2001	1,002	19,514	45,740
Number of informal houses, 2011	2,536	23,138	59,268
Number of informal houses, 2016	2,327	34,845	76,062
Percentage of houses informal, 1996	13.1	23.2	38.2
Percentage of houses informal, 2001	14.7	26.0	40.3
Percentage of houses informal, 2011	25.8	19.3	29.8
Percentage of houses informal, 2016	19.7	23.2	29.0
Presence of traditional land areas	No	No	Yes

made a considerable investment in a new mine, Kolomela. This required the expansion of the town of Postmasburg. Kumba mainly provided rental housing to its employees. A separate organisation, the Tsassamba committee, was created to ensure the adequate expansion of the town and manage the mines' and the municipality's investments.

These mining expansions contributed to population and economic growth. The population in Postmasburg nearly doubled from 19,000 in 1996 to 35,000 in 2015. The Tsantsabane Local Municipality figures show an increase from 26,000 in 1996 to just under 40,000 in 2016. The number of informal houses increased from 800 in 1996 to 2,300 in 2016. Mining expansion brought not only new mineworkers to the town but also a flood of job seekers, with the resulting growth in informal settlements. The number of mineworkers increased rapidly, from 1,200 in 1996 to 7,900 in 2019. In 1996, mining constituted 17% of Postmasburg's economy; by 2019 it had grown to 53%. This was the result of a 2.44% per annum growth in the mining sector between 1996 and 2016, boosting the town's economic growth by 3.55% over that period.

In line with government policy, Beeshoek was normalised by closing it and relocating the people to Postmasburg. The relocation from the company town gave the Beeshoek mineworkers access to homeownership in Postmasburg – this was part of an incentive to relocate, along with a substantial subsidy from the mining company.

China is the town's primary export market for iron ore – and its main economic risk. The global financial crises of 2007/2008 did not affect production. Between 2007 and 2013, the town's mining sector grew by 3.72% per annum and its local economic growth by more than 5% per annum. But more recently, there has been evidence of economic vulnerability. After reaching a record high of USD 168 per metric ton in 2011, prices slumped to USD 55 in 2015. Although prices recovered to USD 109 in 2020, the slump indicated the vulnerability of the iron ore prices. During 2015 and 2016, several small businesses closed in Postmasburg and mineworkers on contract were the first to lose their jobs. The slump in 2015/2016 is proof of the area's vulnerability because of its dependence on the Chinese economy.

Further information on Tsantsabane and Postmasburg can be found in *Mining and Community in South Africa: From Mining Town to Iron Town*, edited by Marais, Burger and Van Rooyen (2018a).

1.6.2 Emalahleni

The urban area of Emalahleni Local Municipality (formerly known as Witbank) was established in 1903, about 50 km east of the border of what is today Gauteng – South Africa's economic heartland. The abundant coal resources were central to the development of the town, and the rail link between Witbank and the gold mines on the Witwatersrand increased its prosperity. In the early 1930s, the government established the first coal-driven power station. Further downstream, activities occurred when Anglo American established a steel factory in the 1960s (closed down in 2013). Since the mid-1990s, coal production has expanded, because of considerable economic growth in South Africa after the democratic transition, the global demand for coal between 2000 and 2007 and the South African rand's depreciation against the US dollar. South African mining companies export approximately 30% of their coal and sell most of the remaining 70% to Eskom (the state-owned electricity utility). The relatively high export price pushed up the internal price. Some of the companies found it worthwhile to export all their coal and pay the penalties for not delivering coal to Eskom.

The increased demand for coal resulted in a boom in the local economy. Emalahleni's average annual economic growth stood at 2.66% per annum from 1996 to 2019, and between 2001 and 2007 it reached 3.04% per annum. The global financial crises did not significantly affect coal production, and consequently the local economy was unaffected. Emalahleni's economy was still experiencing a boom in 2018 because of the high international coal price, although there were signs of decline. The population increased rapidly at 3.34% per annum between 1996 and 2016.

One of the most significant normalisation processes took place in Emalahleni during the 1990s (Van der Watt & Marais, 2019). By the late 1980s, many of the mines had small settlements surrounding them, consisting of family housing for the white workforce, mostly rented from the company, and single-sex hostels for male black workers. However, the mines decided to get rid of both. By 2020, most mining companies in South Africa had closed their mining villages, and the single-sex hostels in the coal industry were the first to be dismantled – some of them as far back as early 1990 (Cloete & Marais, 2021). The unions and the government played a crucial role in this. For black mineworkers, it meant a rapid increase in their monthly salaries as they too were now eligible for living-out allowances. This affected the urban area of Emalahleni as mineworkers from the single-sex hostels or the mining villages were now looking for housing in the town. The rapid increase in informal settlements and the municipality's inability to cope with the increased pressure are a direct result of the new mine housing policy (Campbell et al., 2017) and the mining companies' support for the policy by offering living-out allowances.

The government declared many former mining company towns now open towns in which homeownership was possible. This was the case, for example, with the Rietspruit settlement in Emalahleni municipality. As a partnership programme between the mines and the government, the conversion of Rietspruit was initially hailed by researchers as South Africa's best example of normalisation. However, more recently, concern has been expressed about poor maintenance of Rietspruit's public infrastructure, people not wanting homeownership and the lack of the expected social cohesion (Van der Watt & Marais, 2019).

Today, the main urban area of Emalahleni, formerly Witbank, is an open town providing housing to most of the mineworkers in the area. The mines use buses to transport their workers between the town and the mines. The town has grown so rapidly over the past 20 years that the local government has been unable to cope, as is evident from the rapid increase in the numbers living in informal settlements. In 1996, there were only 13,000 informal houses; by 2016, there were approximately 35,000. A relatively small percentage of the inhabitants are mineworkers, the rest being those who hope to benefit indirectly from the activities of mining (see Chapter 7). This growth, coupled with policies of homeownership and open town development, could create long-term dependencies.

Table 1.1 shows that mining contributed 28% of gross value added (GVA) in Emalahleni in 2019 despite various attempts to diversify the economy. Some approaches to diversification may simply be reinforcing the dependence on mining. For example, Eskom has given many small transport businesses (most of them black-owned) the opportunity to transport coal between the mines and the power stations.[3] Businesses of this kind are at severe risk if the mines downscale or close. The same applies to many other downstream businesses linked to the coal industry.

There are several risks associated with the future of coal that will affect mineworkers in Emalahleni. Global demand for coal is declining. The dollar price of South Africa's coal exports has been declining since 2011, despite some growth in 2018, largely because of the global commitment to clean energy. Most countries of the world, including South Africa, have signed the Paris Agreement, which means they will have to reduce their CO_2 emissions. India, one of the leading countries to which South Africa exports coal, has already adjusted its long-term energy-generation programme and will construct fewer coal-fired power stations than initially planned. South Africa's programme for the development of renewable energy is making only slow progress. To comply with the Paris Agreement, it will have to increase its expected energy from renewables rapidly. At the same time, many of the coal-fired power plants in Emalahleni are old and unlikely to be operating for much longer.

There is much scope to think about diversifying the economy and building on the economic links with Gauteng, but concerns remain that the short-term and long-term implications of mine closure and coal-fired power stations becoming dysfunctional have not been accounted for. Emalahleni needs to think carefully about the existing dependencies and find ways to reduce the effects on the town.

Further information on Emalahleni can be found in *Coal and Energy in Emalahleni, South Africa: Considering a Just Transition*, edited by Marais, Burger, Campbell, Denoon-Stevens and Van Rooyen (2022a).

1.6.3 Rustenburg

The town of Rustenburg, established in 1850, received municipal status from the Union of South Africa in 1918. Today, Rustenburg Local Municipality consists of the town of Rustenburg, several small towns, commercial farming areas and areas under traditional rule. Platinum mining began here on a small scale in the 1920s.The establishment of a platinum smelter in Rustenburg in 1969 stabilised the industry (Davenport, 2013). Platinum was initially used primarily for jewellery. Industrial use became common only from the mid-1980s when the automotive industry started using it in catalytic converters, which are used to cut down emissions from cars, in accordance with legislation in the US and Europe.

Rustenburg expanded as a result of the increased demand for platinum. The population grew from 312,000 in 1996 to 626,000 in 2016, a growth rate of 3.55%. During the mining boom between 2001 and 2007, Rustenburg recorded an economic growth of nearly 5% per annum. Since the global financial crises there has been rapid decline. Despite an upsurge in the price of platinum in 2020, mine employment between 2013 and 2019 dropped by nearly 15,000 workers. The population growth has increased the number of informal houses in the area, from 28,000 in 1996 to 76,000 in 2016. Managing this increase in informal houses is particularly complicated on traditional land where very few land use regulations apply and it is difficult to provide services.

Rustenburg developed primarily as an open town (Ntema, 2019). However, some mining companies do own a small number of houses in Rustenburg, which they rent to their employees. Most of the mines had single-sex hostels. Although most have been dismantled or transformed into more appropriate housing, there is evidence that contract mineworkers still tend to live in the remaining hostels (Bezuidenhout & Buhlungu, 2011).[4] One of the consequences of developing as an open town has been the proliferation of informal settlements – mainly on traditional land.

Three main issues have influenced Rustenburg and will continue to do so. The global financial crises of 2007/2008 led to a decline in car sales and thus the demand for platinum, showing how vulnerable the platinum industry and Rustenburg are to global market trends. The shift from fuel-driven to electric cars in response to the pressure to deal with climate change will further affect the platinum industry, though the shift may be gradual as most vehicles are likely to remain hybrids in the medium term. And finally, competition from the open-cast platinum mines to the north and in Zimbabwe will offer competition to Rustenburg's mines, which are mainly underground and more expensive to operate.

Further information on Rustenburg can be found in *Mining and Community in the South African Platinum Belt: A Decade after Marikana*, edited by Marais, Campbell, Denoon-Stevens and Van Rooyen (2021).

1.7 Chapter outlines

The remaining chapters are outlined as follows:

Chapter 2 (Place attachment, asset-based development and social disruption) discusses the book's two theoretical constructs: place attachment and asset-based development. The concepts are discussed within the context of the dependencies that they create. The literature views these constructs mainly as positive features linked to stability in society. When people belong and have assets, their behaviour becomes less risky, contributing to society and creating stability. But both place attachment and asset-based development assume long-term prosperity (and sometimes intergenerational prosperity). Mine decline and closure can disrupt these ideas. The problem is that policy for mine housing does not consider the likelihood of disruption and assumes that place attachment and asset-based development will counter the effects of mine closure.

Chapter 3 (Mine closure and planning for decline) links two sets of literature: on mine closure and on shrinking cities. Although the chapter focuses on the global mine closure literature, attention is devoted to South Africa. The concept of shrinking cities has grown rapidly over the past three decades and provides valuable ideas on how urban areas can plan for shrinking. The chapter contrasts planning for shrinking with ideas about mining as a long-term driver of the economy. There is a simplistic assumption that many mining areas will find an alternative economic base. The shrinking cities literature suggests the opposite, as it emphasises the value of economic decline and the importance of planning for it. This offers lessons for housing and settlement policies.

Chapter 4 (Post-apartheid housing policy and mining towns) analyses post-apartheid housing policy, particularly housing policy associated with the mining. Using ideas from evolutionary governance theory, the chapter first discusses the history of housing and mine housing in South Africa and then assesses post-apartheid policies. The chapter shows how policies that promote place attachment and asset-based development create the required stability but create new dependencies that the mining industry and South African government has not thought through well. Post-apartheid policies are hampered by goal dependency which results in new path dependencies.

Chapter 5 (Post-apartheid migration labour patterns) introduces three empirical studies focusing on migration patterns from the household surveys. The chapter first looks broadly at mining-related migration in post-apartheid South Africa. It then investigates the different migration patterns of mineworkers and non-mineworkers in the three case study areas, Tsantsabane, Emalahleni and Rustenburg. It also looks at the migration of non-mineworkers into these three areas. The main argument in the chapter is that some form of migrant labour still exists, despite the policy approaches that largely ignore this reality.

Chapter 6 (Mining, housing and wealth creation) compares the housing profiles and housing responses in the three case studies. The analysis provides evidence of asset-building and compares the reality on the ground with the policy intention. The three main questions in the chapter are whether housing has created assets and stability, what the migration trends imply for housing and what likelihood there is of disruption if the mines close. The chapter provides an overview of the three case studies' housing profiles and discusses the potential effects of mine closure.

Chapter 7 (Mining and informal housing) profiles the informal settlements in the three case studies. The chapter suggests that the high percentage of informal housing in all three case study areas requires a more detailed assessment of who lives in these houses, why and what functional role informal houses play. The focus is on informal settlements created both by mineworkers and by non-mineworkers who settle in the mining areas. The chapter discusses the institutional responses to informal settlements and the value of informality in mining areas.

Chapter 8 (Findings, contributions and policy recommendations) frames the book's evidence and arguments in the context of the dependencies outlined in Chapter 1. The chapter contributes to the literature on dependencies in evolutionary governance theory and new institutional economics. In addition to this conceptual discussion, the chapter considers several policy options for mine housing in South Africa. The framework for thinking about mine housing beyond mine closure could have implications for mine housing policy not only in South Africa but also beyond South Africa's boundaries.

Notes

1 I am of course not the first to use the QWERTY analogy. I give due recognition, for example, to David's 'economics of QWERTY' (1985).
2 'Place attachment', that is, people's positive feelings about the place where they live, is discussed in detail in Chapter 2.
3 Eskom's main reason for resorting to this is its dire financial situation, with limited funds available for capital investments in conveyor belts or rail transport.
4 We do not have more recent evidence on the numbers in single-sex hostels, but from observation, this still seems to be true.

References

Bainton, N. & Holcombe, S., 2018. A critical review of the social aspects of mine closure. *Resources Policy*, 59, pp. 368–478.

Bezuidenhout, A. & Buhlungu, S., 2011. From compounded to fragmented labour: mineworkers and the demise of compounds in South Africa. *Antipode*, 43(2), pp. 237–263.

Campbell, M., Nel, V. & Mpambukeli, T., 2017. A thriving coal mining city in crisis? The governance and spatial planning challenges at Witbank, South Africa. *Land Use Policy*, 62, pp. 223–231.

Chaloping-March, M., 2017. *Social Terrains of Mine Closure in the Philippines*. London: Routledge.

Cloete, J. & Marais, L., 2021. Mine housing in the South African coalfields: the unforeseen consequences of post-apartheid policy. *Housing Studies*. https://doi.org/10.1080/02673037.2020.1769038.

Coase, R., 1937. The nature of the firm. *Economica*, 4, pp. 386–405.

Coase, R., 1960. The problem of social cost. *Journal of Law and Economics*, 3, pp. 1–44.

Crous, C., Owen, J., Marais, L., Khanyile, S. & Kemp, D., 2021. Public disclosure of mine closures by listed South African mining companies. *Corporate Social Responsibility and Environmental Management*, 28(3), 1032–1042.
Crush, J., 1989. Migrancy and militance: the case of the National Union of Mineworkers of South Africa. *African Affairs*, 88, pp. 5–13.
Crush, J., 1992. The compound in post-apartheid South Africa. *Geographical Review*, 82(4), pp. 388–400.
Crush, J., 1994. Scripting the compound: power and space in the South African mining industry. *Environment and Planning D: Society and Space*, 12(3), pp. 301–324.
Dallas, S., 1988. *Norman: Colorado Ghost Towns and Mining Camps*. Boulder: University of Oklahoma Press.
Davenport, J., 2013. *Digging Deep: A History of Mining in South Africa*. Johannesburg: Jonathan Ball.
David, P., 1985. Clio and the economics of QWERTY. *American Economic Review*, 75, pp. 332–337.
De Vries, M. & Nemec, J., 2013. Public sector reform: an overview of recent literature and research on NPM and alternative paths. *International Journal of Public Sector Management*, 26(1), pp. 4–16.
Deacon, L., Van Assche, K., Papineaua, J. & Gruezmacher, M., 2018. Speculation, planning, and resilience: case studies from resource-based communities in Western Canada. *Futures*, 104, pp. 37–46.
Digby, C., 2012. Mine closure through the 21st century looking glass. In: A. Fourie, ed. *Proceedings of the 7th International Conference on Mine Closure*. Perth: Australian Centre for Geomechanics, pp. 33–38.
Foucault, M., 1982. The subject and power. *Critical Enquiry*, 8(4), pp. 777–792.
Garner, J., 1992. *The Company Town: Architecture and Society in the Early Industrial Age*. Oxford: Oxford University Press.
Gough, K., Yankson, P. & Esson, J., 2018. Migration, housing and attachment in urban gold mining settlements. *Urban Studies*, 56(13), 2670–2686.
Green, H., 2011. *The Company Town: The Industrial Edens and Satanic Mills That Shaped the American Economy*. New York: Basic Books.
Greiff, A., 2006. *Institutions and the Path to a Modern Economy*. Cambridge: Cambridge University Press.
Haney, M. & Shkaratan, M., 2003. *Mine Closure and Its Impact on the Community. Five Years after Mine Closure in Romania, Russia and Ukraine*. Washington, DC: World Bank.
Haslam McKenzie, F., 2016. The socio-economic impacts of long-distance commuting on people and communities. In: F. Haslam McKenzie, ed. *Labour Force Mobility in the Australian Resources Industry: Socio-economic and Regional Impacts*. Singapore: Springer.
Haslam McKenzie, F., 2020. Long distance commuting: a tool to mitigate the impacts of the resources industries boom and bust cycle? *Land Use Policy*, 93, p. 103932.
IIED (International Institute for Environment and Development), 2002. *Breaking New Ground: Mining, Minerals and Sustainable Development*. London: Earthscan.
Jackson, R., 2002. *Capacity Building in Papua New Guinea for Community Maintenance during and after Mine Closure*. London: International Institute for Environment and Development.
Keeling, A. & Sandlos, J., 2015. *Mining and Communities in Northern Canada: History, Politics, and Memory*. Calgary. Alberta: University of Calgary Press.
King, M., 1993. The 'truth' about autopoiesis. *Journal of Law and Society*, 20(2), pp. 218–236.

Klijn, E., 2008. Governance and governance networks in Europe: an assessment of ten years of research on the theme. *Public Management Review*, 10(4), pp. 505–525.
Krasner, S., 1984. Approaches to the state: alternative conceptions and historical dynamics. *Comparative Politics*, 16(2), pp. 223–246.
Laburn-Peart, C., 1990. Homeownership schemes for black mineworkers: overstating and underpplanning. *Urban Forum*, 1, pp. 69–82.
Laurence, D., 2002. Optimising mine closure outcomes for the community – lessons learned. *Mineral & Energy*, 17, pp. 27–34.
Luhman, N., 1995. *Social Systems*. Stanford, CA: Standford University Press.
Mabin, A., 1993. Capital, coal and conflict: the genesis of planning a company town in Indwe. *Contree*, 34, pp. 21–31.
Macmillan, H., 2012. Mining, housing and welfare in South Africa and Zambia: a historical perspective. *Journal of Contemporary African Studies*, 30(4), pp. 539–550.
Manjunath, A., Paul, P. & Paul, B., 2016. Assessment of socio-economic impacts due to mine closure – a conceptual model. *Journal of Mines, Metals and Fuels*, 64(8), pp. 341–347.
Marais, L., 2013a. Mine downscaling in the Free State Goldfields. *Urban Forum*, 34, pp. 503–521.
Marais, L., 2013b. Resources policy and mine closure in South Africa: the case of the Free State Goldfields. *Resources Policy*, 38, pp. 363–372.
Marais, L., Burger, P., Campbell, M., Denoon-Stevens, S. & Van Rooyen, D., 2022a. *Coal and Energy in Emalahleni, South Africa: Considering a Just Transition*. Edinburgh: Edinburgh University Press.
Marais, L., Burger, P. & Van Rooyen, D., 2018a. *Mining and Community in South Africa: From Mining Town to Iron Town*. London: Routledge.
Marais, L., Campbell, M., Denoon-Stevens, S. & Van Rooyen, D., 2021. *Mining and Community in the South African Platinum Belt: A Decade after Marikana*. New York: Nova.
Marais, L. & Cloete, J., 2013. Labour migration, settlement and mine closure in South Africa. *Geography*, 98(2), pp. 77–84.
Marais, L. & De Lange, A., 2021. Anticipating and planning for mine closure in South Africa. *Futures*, 125, p. 102669.
Marais, L., Haslam McKenzie, F., Deacon, L., Nel, E., Van Rooyen, D. & Cloete, J., 2018b. The changing nature of mining towns: reflections from Australia, Canada and South Africa. *Land Use Policy*, 76, pp. 779–788.
Marais, L., Ntema, J., Campbell, M., Cloete, J. & Lenka, M., 2022b. Informal settlements in the mining context. In: L. Marais, P. Burger, M. Campbell, S. P. Denoon-Stevens & D. van Rooyen eds *Coal and Energy in South Africa: Considering a Just Transition*. Edinburgh: Edinburgh University Press, pp. 73–86.
Maturana, H.R. & Varela, F.J., 1972. Autopoiesis and cognition: the realization of the living. In: *Boston Studies in the Philosophy and History of Science*, 1st edn. Dordrecht: Reidel.
McGuire, G., 2003. *Managing Mine Closure Risks in Developing Communities: A Case Study*. Balikpapan: Kelian Equatorial Mining.
Neil, C., Tykkylainen, M. & Bradbury, J., 1992. *Coping with Closure: An International Comparison of Mine Town Experiences*. London: Routledge.
North, D., 1990. *Institutions, Institutional Change and Economic Performance*. Cambridge: Cambridge University Press.
North, D., 2005. *Understanding the Process of Economic Change*. Princeton, NJ: Princeton University Press.

Ntema, J., 2019. Rustenburg: boom and bust in a mining town. In: L. Marais & V. Nel, eds. *Space and Planning in Secondary Cities: Reflections from South Africa*. Bloemfontein: SunMedia, pp. 203–218.
Obeng-Odoom, F., 2009. Oil and urban development in Ghana. *African Review of Economics and Finance*, 1(1), pp. 18–39.
Porteous, J., 1970. The nature of the company town. *Transactions of the Institute of British Geographers*, 51, pp. 127–142.
Rao, P. & Pathak, K., 2009. Impacts of mine closure on the quality of life of the neighbouring community. *Eastern Journal of Psychiatry*, 12, pp. 10–15.
Schumann, S., 2020. Multicultural musical landscapes: investigating musical influences in Southern West Virginia Company Towns. *The Macksey Journal*, 1, p. 189.
Sesele, K., Marais, L., Van Rooyen, D. & Cloete, J., 2021. Mine decline and women: reflections from the Free State Goldfields. *The Extractive Industries and Society*, 8(1), 211–219.
Siyongwana, P., 2018. The socio-economic impacts of mine closure on local communities: evidence from Mpumalanga Province in South Africa. *GeoJournal*, 84, pp. 367–380.
Smith, J., 2011. *Boom Towns & Relic Hunters of Washington State: Exploring Washington's Historic Ghost Towns & Mining Camps*. Las Vegas, NV: Classic Day Publishing.
Strongman, J., 1992. *Mine Closure. An Overview of the Issues*, Paper presented at the Government of Indonesia Mine Closure Workshop, Jakarta, Indonesia.
Tomlinson, M., 1997. *Mortgage Bondage? Financial Institutions and Low-cost Housing Delivery*. Johannesburg: Centre for Policy Studies Research Report.
Ubels, H., Bock, B. & Haartsen, T., 2019. An evolutionary perspective on experimental local governance arrangements with local governments and residents in Dutch rural areas of depopulation. *Environment and Planning C: Politics and Space*, 37(7), pp. 1277–1295.
Van Assche, K., Beunen, R. & Duineveld, M., 2014. *Evolutionary Governance Theory: An Introduction*. Heidelberg: Springer.
Van Assche, K., Beunen, R. & Duineveld, M., 2016. An overview of EGT's main concepts. In: R. Beunen, K. Van Assche & M. Dunineveld, eds. *Evolutionary Governance*. Heidelberg: Springer, pp. 1–6.
Van Assche, K., Gruezmacher, M. & Deacon, L., 2020. Land use tools for tempering boom and bust: strategy and capacity building. *Land Use Policy*, 93, 103994.
Van Assche, K., Gruzmacher, M. & Granzow, M., 2021. From trauma to fantasy and policy: the past in futures of mining communities; the case of Crownest Pass, Alberta. *Resources Policy*, 72, 102050.
Van der Watt, P. & Marais, L., 2019. Normalising mining company towns in Emalahleni, South Africa. *The Extractive Industries and Society*, 6(4), pp. 1205–1214.
Van der Watt, P. & Marais, L., 2021. Implementing social and labour plans in South Africa: reflections on collaborative planning in the mining industry. *Resources Policy*, 71, 101984.
Van Dijk, E. & Vermunt, R., 2000. Strategy and fairness in social decision making: sometimes it pays to be powerless. *Journal of Experimental Social Psychology*, 36(2), pp. 1–25.
Vivoda, V., Kemp, D. & Owen, J., 2019. Regulating the social aspects of mine closure in three Australian states. *Journal of Energy & Natural Resources Law*, 37(4), pp. 405–424.
Von Krogh, G. & Roos, J., 1995. *Managing Knowledge: Perspectives on Cooperation and Competition*. London: Sage.
Warhurst, A., Macfarlane, M. & Wood, G., 1999. Planning for mine closure: socio-economic impacts. *Minerals & Energy – Raw Materials Report*, 14(3), pp. 21–26.

Watson, I. & Olalade, M., 2019. The state of mine closure in South Africa – what the numbers say. *Journal of the Southern African Institute of Mining and Metallurgy*, 119(7), pp. 639–645.
Williamson, O., 1975, *Markets and Hierarchies: Analysis and Antitrust Implications*. New York: Free Press.
World Bank, 2002. *It's Not Over When It's Over: Mine Closure around the World*. Washington, DC: World Bank Group Mining Department.
World Bank, 2018. *Managing Coal Mine Closure: Achieving a Just Transition for All*. Washington, DC: World Bank.

2 Place attachment, asset-based development and social disruption

2.1 Introduction

I started this book by taking the QWERTY keyboard as an example of how we can be stuck with the consequences of historical decisions. Mining booms create long-term dependencies for the mining company, the community and the mineworkers. Mine closure is a critical juncture that threatens to disrupt these dependencies. The scale of the disruption will depend to some extent on the strength of the dependencies. If care is taken during the boom to minimise the dependencies, managing mine closure will be easier. North (2005) calls these long-term dependencies 'lock-ins'.

Two concepts entrenched in policy and practice contribute to these dependencies: place attachment and asset-based development. Generally, the literature views these concepts as positives. However, in a mining town they may lead eventually to lock-ins. I argue that to evaluate these concepts, we need to look at the context. In the mining context, these concepts should not be applied blindly. Applying them as unquestioned principles is likely to create a false sense of prosperity and increase disruption when mines close. Policies of place attachment and asset-based development emphasise stability and continuity. But mine closure is a major change, and social disruption may be the result. The aim of policy should be to minimise disruption for everyone concerned.

Despite the literature's emphasis on the positive nature of place attachment and asset-based development, these two concepts, and their link with social disruption, require closer scrutiny. In this chapter, I look critically at these concepts, which are central to the current direction of housing policy in South Africa and many other parts of the world. I highlight the problems associated with these concepts and consider their potential for causing social disruption.

2.2 Place attachment

The term 'place attachment' refers to the relationship between people and places, the bond between people and places and the way environments affect behaviour

DOI: 10.4324/9781003262015-2

(Morgan, 2010). Research generally sees place attachment as a positive, like the bond between parents and children, and as a prerequisite for local development. Lewicka (2011, p. 307) says that despite mobility and globalisation, place continues to be 'an object of strong attachments'. Interest in people-place relations is growing. Place attachment research is multidisciplinary: psychology, geography, anthropology, sociology, demography, urban studies, gerontology, tourism studies, forestry, architecture, mining-related studies, economics and urban and regional planning have all shown an interest. Geographers, anthropologists and psychologists have been the most prominent contributors. I drew on a wide range of sources for this section, but most particularly on the seminal collection by Altman and Low (1992), the recent book by Manzo and Devine-Wright (2020) and several review papers and book chapters: Lalli (1992), Giuliani and Feldman (1993), Bott et al. (2003), Giuliani (2003), Twigger-Ross et al. (2003), Easthope (2004), Patterson and Williams (2005), Gustafson (2006), Nicotera (2007), Trentelman (2009), Scannell and Gifford (2010), Lewicka (2011), Bonaiuto et al. (2016) and Hernández et al. (2020).

2.2.1 Research history and definitions

The original place attachment research came from the phenomenologist Bachelard (1964), but several others contributed, such as Buttimer and Seamon (1980) and Kohak (1984). Most of this original work was positivistic and excluded qualitative experiences of place attachment. The phenomenological approach did not consider diverse meanings of the term 'place attachment' but assumed consensus about its meaning.

The initial uptake of place attachment research was slow. A research emphasis on change and migration rather than stability and attachment inhibited the development of the field. In addition, the dominant work from psychologists focusing on cognitive aspects and the individual meant that the concept did not attract much attention from other disciplines. The ahistorical nature of place studies was a further hindrance, while the focus in the 1970s was on cultural differences rather than cultural change. Finally, except for geographers, the research did not consider 'place' as a critical concept.

A more eclectic approach has developed since the mid-1980s. Psychologists and anthropologists in particular helped broaden the scope and develop conceptual frameworks. Research from this period questioned the uniform meaning of the term, increased the rigour of definition and developed taxonomies of concepts. By the 1990s, place attachment research had become more theoretical and the concept began to appear helpful for dealing with practical problems. Altman and Low (1992, p. 4) saw it as an 'integrating concept comprising interrelated and inseparable aspects' whose 'various and complex' origins could contribute to 'individual, group and cultural self-definition and integrity'. Despite theoretical progress, a review in 2011 noted that individual levels of place attachment had dominated research at the expense of place and process components (Lewicka, 2011). This inhibited the theoretical development of place attachment studies.

Stokols and Shumaker (1981) provided the first broad definition of place attachment, which emphasised place dependence and noted that people compare places and decide accordingly. Since then, definitions have varied, but most researchers see it as a multifaceted concept that characterises the 'bonding between individuals and their important places' (Scannell & Gifford, 2010, p. 1). Consequently, alternative terms have been used, such as 'place identity', 'insideness', 'environmental embeddedness', 'rootedness', 'community identity' and 'topophilia' (for a particularly strong love of place).

The definitions have several commonalities (Altman & Low, 1992; Kyle et al., 2004; Scannell & Gifford, 2010; Lin & Lockwood, 2014; Shaykh-Baygloo, 2020). 'Attachment' is seen as meaning an emotional reaction (affect), but also involving cognition and behaviour. This emotional reaction is about love for a place or home, happiness, satisfaction, emotional embeddedness and wellbeing. These emotions usually go with thought, knowledge, beliefs (cognition) and behaviour. Defining the word 'place' is more difficult. It refers to the space for which people provide meaning. Definitions of place differ because of variation in scale, specificity and tangibility. The 'place' that has created attachment can be very big (the world), medium-sized (cities or neighbourhoods), small (homes) or very small (a tree). And it can be imaginary or intangible. The definitions usually note the actors, whether individual, community or a cultural group. The relationship itself is defined: although most place attachment studies link the attachment to the environment, it might also involve social relations (family, friends) in a specific place. Finally, the issue of time is important. In practice, place attachment can be temporal, linear or cyclic: it can fade away, increase or go through phases of growth and decline.

Four factors contribute to the formation and upkeep of place attachment: biological, environmental, psychological and sociocultural (Altman & Low, 1992; Brown et al., 2003). Biological attachment refers to how humans adapt to the physical environment. Territoriality is a common example. Human evolution lies at the centre of this 'fit'. Although people are seldom consciously aware of this type of attachment, it is the basis of many people-environment interactions. The most important proof of this link comes in studies showing how the physical environment affects human behaviour. Research refers to this as environmental determinism. Psychological place attachment depends on individual experiences and events, and researchers use childhood experiences as examples. Sociocultural place attachment depends on norms, attitudes and ideology.

There are four related dimensions of place attachment: place dependence, place identity, affective attachment and social attachment (Raymond et al., 2010; McCunn & Gifford, 2014; Anton & Lawrence, 2016; Shaykh-Baygloo, 2020). Place dependence refers to people's functional goals, which depend on whether one place provides for their needs better than other places. Satisfactory levels of place dependence are likely to result in people choosing to stay in a specific place. Place identity refers to 'those dimensions of self that define the individual's identity concerning the physical environment using a complex pattern of conscious and unconscious ideas, beliefs, preferences, feelings, values, goals, and behavioural tendencies and skills relevant to this environment' (Proshansky, 1978, p. 155). It is usually a symbolic attachment to place, but it provides meaning, creates belonging and helps with self-definition (either individually or for communities). It helps to create feelings of distinctiveness, continuity, self-esteem and self-efficacy. Affective attachment refers to the emotional component of attachment associated with place attachment. People develop emotional relationships (positive or negative) with places. The positive ones contribute to place attachment. Finally, social attachment refers to the bonds between people in a specific place.

The notion of place dependence is central to the argument in this book, although it is related to the other three dimensions. This dependence could inhibit mobility and create lock-ins – something from which it is difficult to escape.

2.2.2 *What is the value of place attachment?*

As said, most research views place attachment as a positive phenomenon that promotes stability, survival, a sense of belonging, self-esteem and personal aspirations (Brown & Perkins, 1992; Scannell & Gifford, 2010). It has several psychological benefits (Altman & Low, 1992; Anton & Lawrence, 2016; Shaykh-Baygloo, 2020). These benefits are usually at the individual level. It can provide both security (through stability) and stimulation. The predictability and the opportunities that come with place attachment give people a measure of control. Personal autonomy helps to create individual identity, self-esteem and pride. The same could apply to groups or communities. The benefits of place attachment extend to social interactions: links with family and friends are important, as they bond through symbols and shared experiences. Once again, the benefit is social stability through social relations. Cultural links associated with place attachment help to build this stability at the community level. All of these factors support the development and maintenance of individual, group and cultural identity, which in turn fosters self-esteem.

Besides these psychological benefits of place attachment, the literature points to other benefits (Lewicka, 2005; Yuksel et al., 2010; López-Mosquera & Sánchez, 2013; Shaykh-Baygloo, 2020). For example, place attachment can inspire positive civic action. This might include joining sports or cultural clubs that encourage civic engagement to deal with community problems. Community crime-fighting is another example. But the benefit goes deeper to include responsible behaviour and preserving places of historical significance. Place attachment creates place loyalty.

In an increasingly globalised world, places and their meanings matter to people (Lewicka, 2011). A study by Gustafson (2009) found that the bonding between people and place had increased in the past three decades, despite variation in scale (home, neighbourhood, city or country, and including non-residential places like holiday towns) and despite increased migration.

Some concerns have been expressed about theory and methods in place attachment research (Pretty et al., 2003; Stedman, 2003; Patterson & Williams, 2005; Morgan, 2010; Lewicka, 2011). The relationships between place-related concepts are sometimes unclear, and the research gets stuck on aspects of definitions. Too often, disagreements about definitions hamper theoretical development. The sharp division between qualitative and quantitative studies inhibits coherent responses to the concerns. Very few studies combine these methods. The most noticeable disciplinary divide is between anthropologists and geographers rooted in qualitative research and psychologists devoted to quantitative methods. Finally, the researchers' diverse backgrounds, disciplines and ideologies prevent place attachment research from developing coherently.

Place attachment research in South Africa is sparse. Dlamini et al. (2021) evaluated the nature and scale of place attachment research in South Africa, identifying several themes: belonging, social identity and conversational practice, cultural symbolism and group identity, and inclusion and exclusion. A small but growing body of work has investigated place attachment in mining towns (Ntema et al., 2017; Marais et al., 2018a; Marais et al., 2018b; Marais et al., 2021). Research on this topic is surprisingly sparse, given the mining industry's expansion to so many small places over the past two to three decades.

2.2.3 *What makes people become attached to places?*

The literature points to several factors contributing to place attachment, but the findings are often contradictory and many studies fail to define the term or state the scale of the analysis. Local context remains important for understanding research results. Below is a summary of the main findings from the following sources: Kasarda and Janowitz (1974), Altman and Low (1992), Fried (2000), Brown et al. (2004), Lewicka (2005), Devine-Wright and Howes (2010), Anton and Carmen (2010), Kamalipour et al. (2012), Lestari and Sumabrata (2018), Manzo (2014) and Akinjokun et al. (2018).

Length of stay can be significant. Place attachment takes time to develop and often the longer people stay in a specific place, the more attached they become. Many empirical research projects have investigated the role of socio-economic factors in creating place attachment. Some found that the important factors are older people and higher education levels, others that they are lower socio-economic status and low income. Settlement size and attributes play a role in place attachment. Smaller and isolated communities usually have much higher attachment levels, and access to basic services increases attachment. Homeownership increases place attachment. A strong relationship has been found between social networks and place attachment. Local social bonds and social capital contribute extensively to the development of place attachment. Finally, environmental factors contribute, with scenic rural areas usually having higher levels of place attachment than rundown urban areas.

2.2.4 *Can place attachment be a bad thing?*

As pointed out earlier, place attachment research usually emphasises the positive effects of place attachment. However, several studies have noted possible negative effects (Brown et al., 2004; Kyle et al., 2004; Billig, 2006; Lewicka, 2011; Shaykh-Baygloo, 2020). Place attachment could inhibit migration or mobility because of place dependence. For example, residents could decide to continue living in an area despite hostile conditions simply because they have a history of living in that area or are extremely dependent on it. In some cases, people continue to live in risky crime-ridden areas and ignore the potential danger. Other typical examples are people continuing to live below flood lines despite the risk of flooding. Because change is usually slow, people tend to overlook the possibility of disruption. This is the reason why people ignore warnings about climate change. Unlike the effects of a flood or an earthquake, the immediate implications are not visible. Someone living on a small, low-lying island is unlikely to realise the long-term and intergenerational implications of rising sea levels. In this book, I acknowledge the value of place attachment, yet use it as an example of dependence. This dependence makes people in mining towns ignore the slow changes and refuse to imagine the abrupt change that mine closure will bring.

2.3 Asset-based development and housing

In the early 1980s, concerns were expressed about using income and economic indicators to describe development. Too often, researchers focused on averages, which hid inequalities, and assumed the existence of a normal Bell Curve, which was seldom present. The challenge to these indicators led to alternative understandings of development, one of which was the capabilities approach proposed by Sen (1992, 1993) and developed further by Nussbaum (2002). Many researchers have used these ideas to promote asset-based development and sustainable livelihood approaches. These changes have shifted the emphasis from needs to assets and encouraged the use of multiple indicators to explain and measure development.

2.3.1 *Research history and definitions*

Sen's capabilities approach laid the foundation for a rethink of development and its measurements. In practice, this approach contributed to asset-based development and sustainable livelihoods frameworks. Moser (2006, p. 5) defines assets as 'a stock of financial, human, natural or social resources that can be acquired, developed, improved and transferred across generations', and notes that this stock 'generates flows or consumption well as additional stock'. Such resources are generally described in the literature as five types of *capital*: natural, physical, financial, human and social (see, for example, Roseland, 2000; Goodwin, 2003). *Natural capital* refers to assets in the natural environment, such as land, forests and wetlands, that provide benefits through ecosystem services, *physical capital* to material and engineered infrastructure such as highways and manufactured equipment such as computers, *financial capital* to the role of money or wealth in facilitating productivity, *human capital* to human health, skills and knowledge and *social capital* to the social connections between people that contribute to collective action.

Asset-based development rose to prominence not only because of the influences of Sen and later Nussbaum but also because of criticism of income-based welfare systems. Asset-based development is premised on the assumption that increased income will not lift people out of poverty. Simply stated, helping the poor to spend more is unlikely to achieve this. (Though there are good reasons to increase the spending capacity of the poor in cases where hunger prevails or where vulnerable groups like children and women are neglected.) Proponents of asset-based development say that, unlike income-based welfare, the building of assets could help people escape poverty. They emphasise that this is likely to take place over generations. The argument is that assets create pathways out of poverty (Sherraden, 1991) and give poor people agency (Dietz & Haurin, 2003), which will help them find their way out of poverty and avoid dependence on the state.

In the US, asset-based development integrates housing, education and savings. It has been slow to take off in the Global South and was initially applied largely in rural areas through livelihood approaches (Dani & Moser, 2008), but application in urban areas is increasing (Lloyd-Jones & Rakodi, 2002). In the Global South, it focuses on risk mitigation and prevention and the creation of social safety nets (Moser, 2006).

2.3.2 The value of policy and programmes

Many researchers have contributed to the discussion about the value of the asset-based development approach. I draw on the following in this section: Moser (1996, 1998), De Soto (2000), Elsinga and Hoekstra (2005), Ford Foundation (2006), Moser (2006, 2007), Moser and Felton (2007), Boonyabancha (2008), Moser and Dani (2008), Parkinson et al. (2009), Toussaint and Elsinga (2009), Morgan (2010), Smith (2010), Duncan and Costantino (2011), Marsh and Gibb (2011), McGee (2012) and Clarke (2013).

Asset-based development approaches link housing, wealth creation and poverty alleviation. Although programmes in the US and UK go far beyond housing assets, these are usually the largest single household asset. Various programmes in the US and UK drive the implementation of asset-based development, typically focusing on homeownership, savings and education. Researchers describe the drive towards homeownership as one of the major social changes of the 20th century. For some, it is about the dominance of neoliberalism and the state withdrawing from housing. Others have questioned whether neoliberalism necessarily leads to state withdrawal. However, the global financial crises of 2007/2008 had profound consequences for housing and homeownership programmes, most notably the decline in the number of people who own their houses in the US.

Research has highlighted the social value of asset-based development, saying it improves wellbeing and reduces risky behaviour. However, the direction of causality has been challenged. For example, are the benefits the result of the assets or are the assets the result of inherent individual and community attributes that discourage risky behaviour?

There is a tendency in the Global South to link the notion of housing assets to titling programmes. Although such a link does exist, it is often assumed that it is the only link. De Soto's work emphasises turning dead capital (land without title) into assets by providing legal title. The World Bank has supported this, seeing the title as allowing households to acquire housing finance, improve their houses and participate in the secondary housing market. This approach has been criticised for several reasons, the most common being that it overemphasises one aspect of asset-building, as a silver bullet for development, and assumes that everyone wants to access credit and climb the housing ladder.

Informal settlement upgrading programmes stress the importance of legitimate ownership of a piece of land. This will safeguard households against forced removals and serve as a secure base from which to work, or go out and find work, and improve their education. In addition to assuming a link between housing and poverty alleviation, proponents of asset-based development say homeownership contributes to physical and mental wellbeing and has substantial psychological benefits. For example, good health promotes labour-market participation, particularly at the lower end of the market. The psychological benefits are linked to those of place attachment and include freedom, control and safety. In the long run, these benefits create social stability, reduce risky behaviour and ensure long-term planning.

Housing assets can be used to generate income. A homeowner can rent out a room or use a room for business purposes. Housing improvements, such as renovations or adding rooms, are an excellent example of the slow process of asset accumulation.

Asset-based development has become a common approach to development. Some governments support this approach in parallel to income support programmes. Like place attachment, it promotes stability and continuity and has several psychological advantages. Most of the criticism of asset-based development come from ideological opposition. These criticisms are not my concern in this book. Rather, I question the short-sighted use of this approach in areas whose long-term economic stability is doubtful because of looming mine closure.

2.3.3 *Asset-based development and mining research*

I could not find research that links mining and asset-based development of mineworkers directly. Yet, there are several ways in which mining policies over the last few decades have indirectly supported asset-based development (Carrington et al., 2010; Marais et al., 2018c; Deacon et al., 2018; Van der Watt & Marais, 2019).

The shift from company towns to open towns and normalisation is one example. In the company town model, mineworkers rented their housing from the mining companies. As a result, the company dominated the business and social infrastructure of these towns. For example, the mine was the decision-maker in business development and sports clubs belong to the mine. Normalisation and the development of open towns shifted these responsibilities from the companies to individual households, local governments and local organisations. This shift started in the mid-1980s and became a dominant approach by the 2000s. The change has been welcomed by the government and the mineworkers, and by liberal and more radical scholars. The immediate improvement was that it gave mineworkers the opportunity to develop housing assets and make their own decisions. It also loosened the mining companies' grip on the town and public life. For many radical scholars, minimising the power of multinational mining companies was a central argument for the new approach. Overall, it transferred the long-term responsibility for dealing with mine closure from mining companies to households and the government.

Along with this change in the way of managing company towns, many mining companies over the past 30 years have opted for clean wages. Historically, many of these companies had substantial investments in non-core activities like housing. Thus, the drive towards clean wages, which went hand in hand with higher wages, was an attempt to clean up the complexities in the companies' wage bills and remove fringe benefits like housing. The drive was partly a response to shareholders not wanting to see too many cost factors but also an attempt to manage decline and closure. Fly-in-fly-out arrangements were another way companies could reduce asset-building in mining areas. This system allowed mineworkers to choose where to locate their housing assets. Many preferred to create these assets in large towns rather than remote mining towns.

Another example of indirect support for asset-building is of course a mining boom. Mining booms are very good periods for asset accumulation. Mineworkers usually get plentiful overtime pay and can invest in various types of assets or diversify their assets. The downside is that a boom can create dependencies and hide the likelihood of possible decline.

Although there are exceptions, the message from mining companies over the past four decades has been clear: 'We shall pay you more, but we shall not be responsible for your housing or developing other assets. Please decide for yourself which assets you need and how you will manage them'. There is, of course, much value in creating this level of independence. However, mining companies are less successful in communicating the vulnerability associated with their operations. And many cannot predict resource declines. The development of a clean wage means that mining companies need not plan for the housing consequences of closure. This has now become the responsibility of mineworker households, who are even less able than the mining companies to predict the future.

2.4 Social disruption

I could not find much research linking the topics of place attachment and resource-based communities. For example, a Scopus search using 'place attachment' and 'mining' as keywords produced a list of 18 sources (10 June 2020). However, the topic of social disruption in resource-based communities is prominent. A Scopus search using 'disruption' and 'mining' produced 946 sources (10 June 2020). Furthermore, Altman and Low's (1992) edited collection has only one chapter on the disruption of place attachment. The latest edited collection by Manzo and Devine-Wright (2020) has no chapter on social disruption and the word 'disruption' does not appear once in the book. Overall, the emphasis is usually on social disruption without describing existing place attachment and its contributing factors. In contrast, place attachment research seldom engages with the disruption of place attachment.

In the literature, the term 'social disruption' refers to the negative social consequences of large developments like mining and energy generation. I expand the concept of social disruption by seeing also as a consequence of *decline*. Mine closure is a disrupter of both place attachment and asset-based development.

2.4.1 Research history and definitions

The concept of social disruption first appeared in Shaw and McKay's book *Juvenile delinquency and urban areas* (1942). The term they used was 'social disorganisation'. In analysing delinquency in Chicago, they were the first to identify a direct link between crime and the social attributes of a place. However, the term did not receive much attention until the 1970s, when researchers started using it to explain the way large mining and energy projects in small and remote towns affected local culture negatively and increased mental health problems and crime rates. Brown and Perkins (1992) list several characteristics of social disruption: it affects both individuals and communities, it usually develops slowly, people find its effects overwhelming, it threatens self-definition and it requires a holistic understanding.

Disruption means change, or an interruption of stability, and thus stands in contrast to place attachment and asset-based development, which mean stability and continuity. Stability and attachment and change and disruption can come in cycles. England and Albrecht (1984, p. 234) define social disruption as 'a period of generalised crisis and loss of traditional routines and attitudes', disrupting people's mental health, worldview and social networks. They say it 'strikes at the organisational level where existing businesses and associations must struggle to meet the challenge of newcomers' and 'reaches the community level as the homogeneous culture is disrupted'. They developed this description of social disruption specifically in the context of developments that brought new people into existing settlements. Brown and Perkins (1992, p. 284) refer to 'predominantly negative connections' and alienation because of 'noticeable changes in the people, processes, or places'. The loss of existing routines or the attitude of a specific community explains the changes. This view takes into account the potential changes in people, processes or place, but most research takes only place into account. Ten Brinke et al. (2017) define social disruption as a process that disturbs the normal functioning of society. Essentially, disruption challenges the continuity and stability in the functioning of a community that place attachment creates. However, social disruption is not necessarily just because of newcomers, as England and Albrecht's description suggests. In this book, I consider broader causes.

2.4.2 Causes and effects of social disruption

For the discussion in this section, I relied on the following sources: Brown (1985), Brown and Perkins (1992), Nevel (1995), Milligan (2003), Curry et al. (2010), Park and Stokowski (2009), Ten Brinke et al. (2017), Zhaoa et al. (2020), Boston et al. (2021) and Castro (2021). Typical causes of social disruption in the literature are crime, disasters and environmental change. These can cause involuntary relocations, but voluntary relocations can also cause social disruption. The literature tends to focus on disruption that happens because of a change in place. But disruption can happen because of changes in social relations. Typical examples are divorce, children leaving home, poor health or a death in the family. Social disruption usually leads to mental health difficulties, but research has also noted physical health concerns.

Involuntary relocations offer the best examples of disruption of place attachment. This type of relocation results from large-scale infrastructure projects, such as building dams, that displace people. Such displacements disrupt economic activity and the social relations that existed in the original settlement. One recent example has been the relocation of households affected by dam construction in the Lesotho Highlands Water Project. But these relocations can also happen in urban settings where people need to make way for large projects. The notorious resettlements under apartheid come to mind and Operation Restore Order (2005) is a recent urban example from Zimbabwe. Displacements because of conflicts also fall in this category. Although some involuntary relocations do follow lengthy processes, they are more usually rapid.

Neighbourhood change, sometimes associated with increased crime, is a slow process of disruption. Crime damages the bonding process between people and places. Burglaries, for example, disrupt the ideas of safety, self-definition and identity that are associated with home. Increased crime also affects the people's feelings about their community and could make them relocate.

Disasters such as floods and earthquakes are a major cause of social disruption. They affect place attachment, continuity and assets in several ways. People may be relocated and lose their homes. At a slower pace, disruption can be caused by climate change. People may have to relocate, for example, because of coastal flooding or erosion. Small and low-lying islands and large parts of the Netherlands are at risk. This change is slow and even disputed in some cases. The possibility requires long-term planning, but many governments are more concerned with short-term planning for disasters. It is much easier to support disrupted communities with short-term relief than try to avoid such disasters in the first place.

In this book, I am especially interested in understanding change as a slow process. Quick change in the form of a disaster usually elicits a rapid response from governments. Understanding slow change and the dependencies created in an existing context is far more difficult. For example, governments tend to pay attention to the rapid change that mine developments bring, while ignoring the longer-term process of mine decline, and thus find themselves unprepared for mine closure, which could be abrupt.

2.4.3 *Mining and social disruption*

There is a long history of research linking mining and social disruption. However, the research mainly associates social disruption with mining growth or boomtowns. Mining decline has not been positively linked to social disruption and empirical evidence from the Global North has shown little evidence that such a link exists (O'Connor & Ruddell, 2021). However, Marais et al. (2022) have argued that mine closure should be viewed as a form of social disruption.

Definitions of a boomtown include rapid population growth in mainly rural or smaller settlements because of the discovery of natural resources (Jones & Mayzer, 2021). Boomtowns have four main characteristics: high population increase, long-time residents leaving the area, a housing shortage and male dominance. In addition, booms are bad for local culture and traditions. Finsterbusch and Freudenburg (2002) identified four phases of boomtown research, and Jones and Mayer (2021) added a fifth phase. The body of research linking mine developments and social disruption is fairly extensive. I draw on the following in the discussion below: Smith et al. (2001), Brown et al. (2005), Haslam McKenzie et al. (2009), Carrington et al. (2010), Ennis et al. (2014) and Haslam McKenzie (2020), and particularly on some that focus specifically on linking mining and crime: Ruddell et al. (2014), Axbard et al. (2016), Jones (2016), Ruddell (2017), Gourley and Madonia (2018), Stretesky and Grimmer (2020), Fan et al. (2021), Marais et al. (2022) and O'Connor and Ruddell (2021). A recent systematic review of the relationship between crime and mining (Stretesky & Grimmer, 2020) is probably the most notable contribution.

The initial phase of research in the 1960s largely emphasised the economic value of mining booms for remote places. However, this research originated from economists and was uncritical. This positive view came under scrutiny in the 1970s (the second phase) when the focus shifted to the negative consequences of resource extraction. In emphasising the negative effects, research started to use the term 'social disruption' to explain the consequences of large economic developments like mining on small towns and rural communities. Researchers highlighted the social implications of rapid population growth for long-time residents' attitudes and behaviour, and commonly used crime data, divorce rates, mental health cases and school dropout numbers to make their point. However, it was not long before others started to question the results from this second phase. Boomtown studies in the early 1980s (the third phase) questioned several outcomes of the 1970s research and pointed out its methodological problems. The main concerns were lack of longitudinal data and simplistic conclusions. Too often, researchers used single indicators, data from one year only, and did not attempt to understand the change holistically. The end of the 1980s (the fourth phase) saw a more nuanced analysis of the adverse social consequences of boomtown developments.

By the 2000s (the fifth phase), the complex relationships between boomtowns and social disruption dominated the research. This research questioned the simplistic relationships between social disruption and population and economic growth, and showed that social disruption fades out over time or re-emerges, has gender dimensions and affects indigenous communities and local housing conditions. It also found that increasingly it is households and communities that must manage the negative effects of social disruption. Yet, this research did not consider the complexities of mine closure as an example of social disruption. Empirical evidence from the Global North has shown that crime rates decline with mine closure. However, research from the Global South challenges this finding and shows how mine decline has contributed to increased crime levels in South Africa (Axbard et al., 2016; Marais et al., 2022). The topic of mine decline and closure will increase the complexity of boomtown research in this fifth phase, and I argue later that it constitutes a sixth phase.

One of the dominant themes in boomtown and social disruption theory in mining areas is the relationship between booms and crime. The literature finds specific crimes associated with mining growth: theft of mining equipment, illegal mining, smuggling and property crimes, driving under the influence of alcohol, carjacking, drug abuse, gangsterism and violent or serious crimes. The general finding has been a clear connection between mining and shale gas developments and increased crime. Yet, very little evidence has been found of increased crime once a mine declines or closes, except for some recent work in South Africa (Axbard et al., 2016; Marais et al., 2022).

A second common theme, particularly relevant to this book, is the social disruption associated with increased housing demand because of population growth. This increased demand pushes up rental and housing prices, making it difficult for long-time residents to rent or buy. The literature points to social concerns like deteriorating housing conditions and 'hot bedding' (use of accommodation in shifts). In the Global South, these housing pressures result in the growth of informal settlements near mining sites.

2.5 Conclusion

This chapter explained the book's three central theoretical constructs: place attachment, asset-based development and social disruption. First, it pointed out

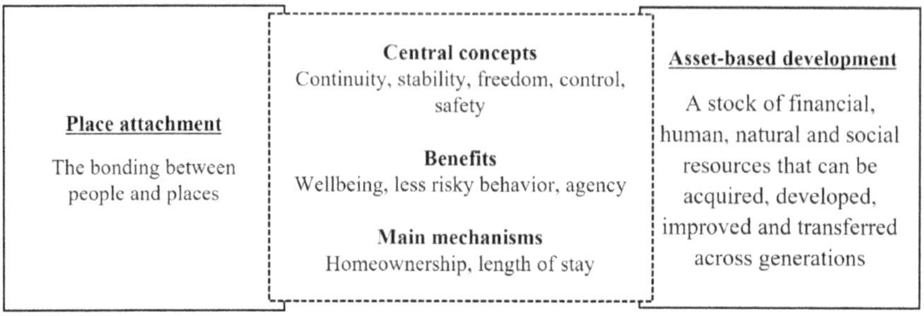

Figure 2.1 A conceptual link between place attachment and asset-based development.

the link between place attachment and asset-based development, despite the origins of these concepts being distinctly different. The stability and continuity associated with place attachment and asset-based development have psychological and wellbeing benefits for individuals, households and communities, encouraging self-esteem and a sense of identity, offering a measure of control, promoting positive civic action and reducing risky behaviour. Neither place attachment research nor asset-based development research has paid enough attention to these links. Linking these two concepts, as shown in Figure 2.1, is one of this chapter's main contributions to the literature.

A second contribution of this chapter is that contrary to the general view expressed in the literature, I do not necessarily view these two concepts in a positive light. Instead, I think both concepts are instrumental in creating dependencies which could be risky in the mining context. As the QWERTY keyboard example illustrates, society can get so used to stability and continuity that change becomes difficult and even impossible. In the case of a mining town, dependence on mining prosperity makes change difficult for households, the community and local organisations. As a result they are unable to respond appropriately.

A third contribution is to the field of boomtown research. Although research from the Global North finds little evidence that mine closure creates social disruption, I argue the opposite in this book. I argue that mine closure needs to be considered as an additional complexity in boomtown research, in effect a sixth phase. Closure in many Global South mining towns is likely to be as disruptive as an earthquake or a flood.

References

Akinjokun, A., Bin Mohd, M., Jusan, M. & Shahminan, R., 2018. Place attachment and determinants of living arrangement in the family house in Ibadan, Nigeria. *African Population Studies*, 32(3), pp. 4559–4573.

Altman, I. & Low, S., 1992. *Place Attachment*. New York: Plenum Press.

Anton, C. & Carmen, L., 2010. Home is where the heart is: the effect of place of residence on place attachment and community participation. *Journal of Environmental Psychology*, 40, pp. 451–461.
Anton, C. & Lawrence, C., 2016. The relationship between place attachment, the theory of planned behaviour and residents' response to place change. *Journal of Environmental Psychology*, 47, pp. 145–154.
Axbard, S., Poulsen, J. & Tolonen, A., 2016. *Extractive Industries, Production Shocks and Criminality: Evidence from a Middle-income Country*. Columbia University, Centre for Economic Development and Policy.
Bachelard, G., 1964. *The Poetics of Space*. New York: Orion.
Billig, M., 2006. Is my home my castle? Place attachment, risk perception, and religious faith. *Environment and Behavior*, 38, pp. 248–265.
Bonaiuto, M., Alves, S., De Dominicis, S. & Petrucelli, I., 2016. Place attachment and natural hazard risk: research review and agenda. *Journal of Environmental Psychology*, 48, pp. 33–53.
Boonyabancha, S., 2008. Upgrading Thailand's urban settlements: a community driven process of social development. In: C. Moser & A. Dani, eds. *Assets, Livelihoods and Social Policy*. Washington, DC: World Bank, pp. 195–214.
Boston, J., Panda, A. & Surminski, S., 2021. Designing a funding framework for the impacts of slow-onset climate change – insights from recent experiences with planned relocation. *Current Opinion in Environmental Sustainability*, 50, pp. 159–168.
Bott, S., Cantrill, J. & Meyers, O., 2003. Place and the promise of conservation psychology. *Human Ecology Review*, 10, pp. 100–112.
Brown, B., 1985. Residential territories: cues to burglary vulnerability. *Journal of Architectural Planning and Research*, 2, pp. 231–243.
Brown, B. & Perkins, D., 1992. Disruption in place attachment. In: I. Altman & S. Low, eds. *Place Attachment*. London: Plenum Press, pp. 279–304.
Brown, B., Perkins, D. & Brown, G., 2003. Place attachment in a revitalising neighborhood: individual and block levels of analysis. *Journal of Environmental Psychology*, 23(3), pp. 259–271.
Brown, B., Perkins, D. & Brown, G., 2004. Incivilities, place attachment and crime: block and individual affects. *Journal of Environmental Psychology*, 24(3), pp. 359–371.
Brown, R., Dorins, F. & Krannich, R., 2005. The boom-bust-recovery cycle: dynamics of change in community satisfaction and social integration in Delta, Utah. *Rural Sociology*, 70(1), pp. 28–49.
Buttimer, A. & Seamon, D., 1980. *The Human Experience of Space and Place*. London: Croon Helm.
Carrington, K., McIntosh, A. & Scott, J., 2010. Globalisation, frontier masculinities and violence: booze, blokes and brawls. *British Journal of Criminology*, 50(3), pp. 393–413.
Castro, D., 2021. Community organisation for the protection of cultural heritage in the aftermath of disasters. *International Journal of Disaster Risk Reduction*, 60, p. 102321.
Clarke, W., 2013. The aftermath of the general financial crises for the homeownership society: what happened to low-income homeowners in the US? *International Journal of Housing Policy*, 13(3), pp. 227–246.
Curry, J., Hanke, M., Piper, M., Bailey, M., Bringardner, B., Sheridan, J. & Marsh, C., 2010. Social disruption induces lung inflammation. *Brain, Behavior, and Immunity*, 22(4), p. 394–402.

Dani, A. & Moser, C., 2008. Asset-based social policy and public action in a polycentric world. In: C. Moser & A. Dani, eds. *Assets, Livelihoods and Social Policy*. Washington, DC: World Bank, pp. 3–41.
De Soto, H., 2000. *The Mystery of Capital. Why Capitalism Triumphs in the West and Fails Everywhere Else*. New York: Basic Books.
Deacon, L., Van Assche, K., Papineaua, J. & Gruezmacher, M., 2018. Speculation, planning, and resilience: case studies from resource-based communities in Western Canada. *Futures*, 104, pp. 37–46.
Devine-Wright, P. & Howes, Y., 2010. Disruption to place attachment and the protection of restorative environments: a wind energy case study. *Journal of Environmental Psychology*, 30, pp. 271–280.
Dietz, R. & Haurin, D., 2003. The social and private micro-level consequences of homeownership. *Journal of Urban Economics*, 54(3), pp. 401–450.
Dlamini, S., Tesfamichael, S. & Mokhele, T., 2021. A review of place identity studies in post-apartheid South Africa. *South African Journal of Psychology*, 51(1), pp. 121–133.
Duncan, D. & Costantino, C., 2011. Effects of the recent credit cycle on homeownership rates across households: what we know and what we expect. In: R. Forrest & N. Yip, eds. *Housing markets and the Global Financial Crises: The Uneven Impacts on Households*. Northampton: Edward Elgar, pp. 20–40.
Easthope, H., 2004. A place called home. *Housing. Theory and Society*, 21, pp. 128–138.
Elsinga, M. & Hoekstra, J., 2005. Homeownership and housing satisfaction. *Journal of Housing and the Built Environment*, 20, pp. 401–424.
England, J. & Albrecht, S., 1984. Boomtowns and social disruption. *Rural Sociology*, 49(2), pp. 230–246.
Ennis, G., Tofa, M. & Finlayson, M., 2014. Open for business but at what cost? Housing issues in boomtown Darwin. *Australian Geographer*, 45(4), pp. 447–464.
Fan, H., Enshaei, H. & Jayasinghe, S., 2021. Safety philosophy and risk analysis methodology for LNG bunkering simultaneous operations (SIMOPs): a literature review. *Safety Science*, 136, p. 105150.
Finsterbusch, K. & Freudenburg, W., 2002. Social impact assessment and technology assessment. In: R. Dunlap & W. Michelson, eds. *Handbook of Environmental Sociology*. Westport, CT: Greenwood Press, pp. 287–299.
Ford Foundation, 2006. *Building Assets to Reduce Poverty and Injustice*. New York: Ford Foundation.
Fried, M., 2000. Continuities and discontinuities of place. *Journal of Environmental Psychology*, 20(3), pp. 193–205.
Giuliani, M., 2003. Theory of attachment and place attachment. In: M. Bonnes, T. Lee & M. Bonaiuto, eds. *Psychological Theories for Environmental Issues*. Aldershot: Ashgate, pp. 137–170.
Giuliani, M. & Feldman, R., 1993. Place attachment in a developmental and cultural context. *Journal of Environmental Psychology*, 13, pp. 267–274.
Goodwin, N., 2003. *Five Kinds of Capital: Useful Concepts for Sustainable Development*. Global Development and Environmental Institute. Working Paper 03–07.
Gourley, P. & Madonia, G., 2018. Resource booms and crime: evidence from oil and gas production in Colorado. *Resource and Energy Economics*, 54, pp. 37–52.
Gustafson, P., 2006. Place attachment and mobility. In: N. McIntyre & K. McHugh, eds. *Multiple Dwelling and Tourism: Negotiating Place, Home and Identity*. London: CAB International, pp. 17–31.
Gustafson, P., 2009. More cosmopolitan, no less local. *European Societies*, 11, pp. 25–27.

Haslam McKenzie, F., 2020. Long distance commuting: a tool to mitigate the impacts of the resources industries boom and bust cycle? *Land Use Policy*, 91, p. 103932.

Haslam McKenzie, F., Phillips, R., Rowley, S., Brereton, D. & Birdsall-Jones, C., 2009. *Housing Market Dynamics in Resource Boom Towns*. Perth: Australian Housing and Urban Research Institute.

Hernández, B., Hidalgo, C. & Ruiz, C., 2020. Theoretical and methodological aspects of research on place attachment. In: L. Manzo, & P. Devine-Wright, eds. *Place Attachment: Advancements in Theory, Methods and Application*. London: Routledge, pp. 125–138.

Jones, F., 2016. Anomie in the oil patch: a preliminary analysis. *International Journal of Rural Criminology*, 3(1), pp. 68–91.

Jones, M. & Mayzer, R., 2021. A fifth wave? The changing nature of boomtown research. *The Extractive Industries and Society*. https://doi.org/10.1016/j.exis.2021.02.013.

Kamalipour, H., Yeganeh, A. & Alalhesabi, M., 2012. Predictors of place attachment in urban residential environments: a residential complex case study. *Procedia – Social and Behavioral Sciences*, 35, pp. 459–467.

Kasarda, J. & Janowitz, M., 1974. Community attachment in mass society. *American Sociological Review*, 39, pp. 328–339.

Kohak, E., 1984. *The Embers and the Stars*. Chicago, IL: University of Chicago Press.

Kyle, G., Mowen, A. & Tarrant, M., 2004. Linking place preferences with place meaning: an examination of the relationship between place motivation and place attachment. *Journal of Environmental Psychology*, 24(4), p. 439–454.

Lalli, M., 1992. Urban-related identity: theory, measurement, and empirical findings. *Journal of Environmental Psychology*, 12, pp. 285–303.

Lestari, W. & Sumabrata, J., 2018. The influencing factors on place attachment in neighbourhood of Kampung Melayu. *IOP Conference Series Earth and Environmental Science*, 126(1), p. 012190.

Lewicka, M., 2005. Ways to make people active: the role of place attachment, cultural capital, and neighborhood ties. *Journal of Environmental Psychology*, 25(4), pp. 381–395.

Lewicka, M., 2011. Place attachment: how far have we come in the last 40 years? *Journal of Environmental Psychology*, 31(3), pp. 207–230.

Lin, C. & Lockwood, M., 2014. Forms and sources of place attachment: evidence from two protected areas. *Geoforum*, 53, pp. 74–81.

Lloyd-Jones, T. & Rakodi, C., 2002. *Urban Livelihoods: A People-Centred Approach to Reducing Poverty*. London: Earthscan.

López-Mosquera, N. & Sánchez, M., 2013. Direct and indirect effects of received benefits and place attachment in willingness to pay and loyalty in suburban natural areas. *Journal of Environmental Psychology*, 34, pp. 27–35.

Manzo, L., 2014. On uncertain ground: being at home in the context of public housing redevelopment. *International Journal of Housing Policy*, 14(4), pp. 389–410.

Manzo, L. & Devine-Wright, P., 2020. *Place Attachment: Advances in Theory, Methods and Applications*. London: Routledge.

Marais, L., Burger, P. & Van Rooyen, D., 2018a. *Mining and Community in South Africa: From Small Town to Iron Town*. London: Routledge.

Marais, L., Cloete, J., van Rooyen, D., Denoon-Stevens, S. & Nel, V., 2018b. Place attachment and social disruption in Postmasburg, a rapidly growing South African mining town. *GeoJournal*, 84(1), 71–83.

Marais, L., Haslam McKenzie, F., Deacon, L., Nel, E., van Rooyen, D. & Cloete, J., 2018c. The changing nature of mining towns: reflections from Australia, Canada and South Africa. *Land Use Policy*, 76, pp. 779–788.

Marais, L., Ndaguba, E., Mmbadi, E., Cloete, J. & Lenka, M., 2022. Mine closure, social disruption and crime in South Africa. *The Geographical Journal*. DOI:10.1111/geoj.12430.

Marais, L., Owen, J., Kotze, T., Nel, P., Cloete, J. & Lenka, M., 2021. Determinants of place attachment among mineworkers: evidence from South Africa. *Extractive Industries and Society*, 8(3), 100943.

Marsh, A. & Gibb, K., 2011. Uncertainty, expectations and behavioural aspects of housing market choices. *Housing, Society and Theory*, 28(3), pp. 215–238.

McCunn, L. & Gifford, R., 2014. Interrelations between sense of place, organisational commitment, and green neighborhoods. *Cities*, 41, pp. 20–29.

McGee, K., 2012. Young people, homeownership and future welfare. *Housing Studies*, 27(6), pp. 853–862.

Milligan, M., 2003. Displacement and identity discontinuity: the role of nostalgia in establishing new identity categories. *Symbolic Interaction*, 26(3), pp. 381–403.

Morgan, P., 2010. Towards a developmental theory of place attachment. *Journal of Environmental Psychology*, 30, pp. 11–20.

Moser, C., 1996. *Confronting Crisis: A Comparative Study of Household Responses to Poverty and Vulnerability in Four Poor Urban Communities*. Washington DC: The World Bank.

Moser, C., 1998. The asset vulnerability framework: reassessing urban poverty reduction strategies. *World Development*, 26, pp. 1–19.

Moser, C., 2006. *Asset-based Approaches to Poverty Reduction in a Globalised World*. Washington, DC: Brookings Institution.

Moser, C., 2007. *Reducing Global Poverty: The Case for Asset Accumulation*. Washington, DC: Brookings Institution Press.

Moser, C. & Dani, A., 2008. *Assets, Livelihoods and Social Policy*. Washington, DC: World Bank.

Moser, C. & Felton, A., 2007. Intergenerational asset accumulation and poverty reduction in Guayaquil, Ecuador, 1978–2004. In: C. Moser, ed. *Reducing Global Poverty: The Case of Asset Accumulation*. Washington, DC: Brooking Institution Press, pp. 18–34.

Nevel, L., 1995. Culture conflict and crime in Alaskan Native villages. *Journal of Criminal Justice*, 23(2), pp. 177–189.

Nicotera, N., 2007. Measuring neighborhood: a conundrum for human services researchers and practitioners. *American Journal of Community Psychology*, 40, pp. 36–51.

North, D.C., 2005. *Understanding the Process of Economic Change*. Princeton, NJ: Princeton University Press.

Ntema, J., Marais, L., Cloete, J. & Lenka, M., 2017. Social disruption, mine closure and housing policy: evidence from the Free State Goldfields, South Africa. *Natural Resources Forum*, 41(1), pp. 31–40.

Nussbaum, M., 2002. Capabilities and social justice. *International Studies Review*, 4(2), pp. 123–135.

O'Connor, C. & Ruddell, P., 2021. After the downturn: perceptions of crime and policing. *The Canadian Geographer*. https://doi.org/10.1111/cag.12671.

Parkinson, S., Searle, B., Smith, S., Stoakes, A. & Wood, G., 2009. Mortgage equity withdrawal in Australia and Britain: towards a welfare state. *International Journal of Housing Policy*, 9(4), pp. 365–389.

Park, M. & Stokowski, P., 2009. Social disruption theory and crime in rural communities: comparisons across three levels of tourism growth. *Tourism Management*, 30(6), pp. 905–915.

Patterson, M. & Williams, D., 2005. Maintaining research traditions on place: diversity of thought and scientific progress. *Journal of Environmental Psychology*, 25, pp. 361–380.

Pretty, G., Chipuer, H. & Bramston, P., 2003. Sense of place amongst adolescents and adults in two rural Australian towns: the discriminating features of place attachment, sense of community and place dependence in relation to place identity. *Journal of Environmental Psychology*, 23, pp. 273–287.

Proshansky, H., 1978. The city and self-identity. *Environment and Behavior*, 10(1), pp. 147–169.

Raymond, C., Brown, G. & Weber, D., 2010. The measurement of place attachment. *Journal of Environmental Psychology*, 30(2), pp. 198–205.

Roseland, M., 2000. Sustainable community development: integrating environmental, economic, and social objectives. *Progress in Planning*, 54, pp. 73–132.

Ruddell, R., 2017. *Oil, Gas and Crime: The Dark Side of the Boomtown*. New York: Palgrave Macmillan.

Ruddell, R., Jayasundara, D., Mayzer, R. & Heitkamp, T., 2014. Drilling down: an examination of the boom-crime relationship in resource based boom communities. *Western Criminology Review*, 15(1), pp. 3–17.

Scannell, L. & Gifford, R., 2010. Defining place attachment: a tripartite organising framework. *Journal of Environmental Psychology*, 30(1), pp. 1–10.

Sen, A., 1992. *Inequality Re-examined*. Oxford: Oxford University Press.

Sen, A., 1993. Capabilities and wellbeing. In: M. Nussbaum & A. Sen, eds. *The Quality of Life*. Oxford: Clarendon Press, pp. 30–53.

Shaw, C. & McKay, H., 1942. *Juvenile Delinquency and Urban Areas*. Chicago, IL: University of Chicago Press.

Shaykh-Baygloo, R., 2020. A multifaceted study of place attachment and its influences on civic involvement and place loyalty in Baharestan new town, Iran. *Cities*, 96, p. 102473.

Sherraden, M., 1991. *Assets and the Poor: A New American Welfare Policy*. New York: M.E. Sharpe.

Smith, D., Krannich, R. & Hunter, L., 2001. Growth, decline, stability, and disruption: a longitudinal analysis of social wellbeing in four western Australian communities. *Rural Sociology*, 66(3), pp. 425–450.

Smith, S., 2010. *Care-full Markets: Miracle or Mirage?* Cambridge: The Tanner Lectures on Human Values, Cambridge University.

Stedman, R., 2003. Is it really just a social construction? The contribution of the physical environment to sense of place. *Society and Natural Resources*, 16, pp. 671–685.

Stokols, D. & Shumaker, S., 1981. People in places. A transactional view of settings. In: J. Harvey, ed. *Cognition, Social Behavior and the Environment*. Hillsdale, NJ: Lawrence Erlbaum Associates, pp. 441–488.

Stretesky, P. & Grimmer, P., 2020. Shale gas development and crime: a review of literature. *The Extractive Industries and Society*, 7(3), pp. 1147–1157.

Ten Brinke, W., Knoop, J., Muilwijk, H. & Ligtvoet, W., 2017. Social disruption by flooding, a European perspective. *International Journal of Disaster Risk Reduction*, 21, pp. 312–322.

Toussaint, J. & Elsinga, M., 2009. Exploring 'housing asset-based welfare'. Can the UK be held up as an example for Europe? *Housing Studies*, 24(5), pp. 669–692.

Trentelman, C., 2009. Place attachment and community attachment: a primer grounded in the lived experience of a community sociologist. *Society and Natural Resources*, 22, pp. 191–210.

Twigger-Ross, C., Bonaiuto, M. & Breakwell, G., 2003. Identity theories and environmental psychology. In: M. Bonnes, T. Lee & M. Bonaiuto, eds. *Psychological Theories for Environmental Issues*. Aldershot: Ashgate, pp. 202–203.

Van der Watt, P. & Marais, L., 2019. Normalising mining company towns in Emalahleni, South Africa. *The Extractive Industries and Society*, 6, pp. 1205–1214.

Yuksel, A., Yuksel, F. & Bilim, Y., 2010. Destination attachment: effects on customer satisfaction and cognitive, affective and conative loyalty. *Tourism Management*, 31(2), pp. 274–284.

Zhaoa, M., Sun, Z. & Zeng, Y., 2020. Exploring urban risk reduction strategy based on spatial statistics and scenario planning. *Journal of Cleaner Production*, 264, p. 121668.

3 Mine closure and planning for decline

3.1 Introduction

Mining growth creates place attachment. People settle down and accumulate assets. They become dependent on the place and unable to imagine life beyond mining. Place attachment and asset-based development assume long-term economic prosperity and continuity, which is a dangerous assumption in a mining town. In these towns, change is more likely than stability. Most research on mining towns has looked at social disruption associated with growth. In this book, I look instead at social disruption caused by mine decline and closure. In Chapter 2, I called this the sixth phase of boomtown research. When a mine closes, it disrupts the place attachment and asset-based development that its growth encouraged. The dependence created by the period of growth threatens to lock people into their town and makes management of closure difficult.

A typical response when a mine is closing is to attempt to diversify the town's economy. But this is difficult in isolated areas and few mining towns have succeeded. I argue that a better option is for the government and mining companies to accept the reality of mine decline. The literature on shrinking cities provides a good framework for planning decline so as to minimise the long-term dependence and liabilities that usually develop because of a mining boom.

This chapter brings together two sets of literature. I first review the literature on mine decline and closure, focusing on discussions of the social consequences and the policy responses. I then review the shrinking cities literature, which, although it comes from outside the mining context, has links with mine closure. It argues that planning for decline is a realistic alternative to planning economic diversification and growth. Planning for shrinking can generate new economic activities, but it is an option that governments and mining companies seldom consider.

3.2 Mine closure

Planning for mine closure is difficult because the two main reasons for closure – the finite nature of mining and the volatility in international commodity prices – make it hard to predict. In addition, the growth-related prosperity that

DOI: 10.4324/9781003262015-3

resource-based communities and mine dependent households enjoy inhibits planning for decline. But the mining industry is under increasing international pressure to show its commitment to sustainable development through responsible mine closure processes.

Before the industrial revolution, mining was a small-scale activity. With the increased need for energy and resources for industry, it became large-scale and capital intensive. Technology and machinery replaced much of the manual labour, boosting production. Resources were depleted faster. Mine closure has become a major concern today. Closures have increased worldwide during the past 50–70 years, and in Africa, with a shorter commercial mining history, in the past 20 years (World Bank, 2002).

Governments have tightened up closure regulations. Up to the 1970s, most mining companies simply abandoned mines when they closed. Left to deal with the environmental and economic consequences, local communities started questioning this practice. Mining companies now have to comply with environmental regulations when closing a mine and rehabilitate the land. But not much attention has been paid to dealing with the social consequences.

Much work has gone into assessing the available mineral deposits worldwide, but not much into assessing the likelihood of mine closures. In an exception, the World Bank predicted in 2002 that 25 large mines in the Global South would close in the next decade – but did not follow up on this or monitor its predictions. Consequently, we know little about future mine closures. The lack of interest is understandable: the economic incentive is to identify sites for new mines, not predict which existing ones will close. And, as noted above, predicting closure is not easy, depending as it does not just on the availability of a mineral, but on the ups and downs of prices.

Several factors make it difficult to plan for closure. The mining companies carry out extensive mine life-cycle planning, based on resource availability, but external factors like resource price complicate their predictions. Planning for closure is also difficult for everyone else involved. As the mining industry seldom passes on its information, the community finds it difficult to plan accordingly. The local government is similarly hampered, and it tends anyway to focus on growth. Asking it to focus simultaneously on growth and decline is asking too much. Mineworkers and households, reaping the benefits of mine employment, find it hard to imagine planning for decline and rather increase their dependency by becoming attached to the place and accumulating assets.

3.2.1 What is mine closure?

Views differ. Emphasising land rehabilitation, Limpitlaw and Briel (2014, p. 1) say mine closure is 'the process of transforming an active mine into a set of safe and stable landforms that are non-polluting and provide habitat and ecosystem services and support economic activities by the new land users'. But for Chaloping-March (2017, p. 2), mine closure is a 'social episode in the ebb and flow of life within communities'. She notes that it is not a straightforward process: it can take many years and involve closures and reopenings. Bainton and Holcombe (2018) refer to mine closure as a process of economic transition. Six main points drawn from the following sources receive attention below: Peck et al. (2005), Stacey et al. (2010), Digby (2012), Wang et al. (2016), Chaloping-March (2017), Bainton and Holcombe (2018) and Vivoda et al. (2019).

First, a mine can close in various ways. Peck et al. (2005) identify five possible closure scenarios: (1) ordered advance and completion, (2) strategic revision of the final mine-life years, (3) ordered retreat, (4) retreat in disarray and (5) dereliction of duty. In scenario 1, closure planning is an integral part of the mining process. It forms part of the preplanning and the mine provides decommissioning guidelines. As I explain later in the chapter, this planning is usually done by the mining company and not by mining communities. In scenario 2, there is adequate time to adjust plans. In scenario 3, the mine is closed prematurely because of internal mining-related factors or a drop in resource prices. This scenario allows for orderly closure because a closure plan is in place. It is more problematic for local communities as it happens unexpectedly and there is not much time to find alternative employment or deal with the expected consequences. Scenario 4 is premature closure without an appropriate closure plan, possibly causing conflict between the affected parties and threats of lawsuits. In this scenario, it is very likely that the financial provisions are insufficient, environmental rehabilitation will not be done and the social aspects will not receive attention. Scenario 5 means that the mining company takes no responsibility for closure, strips the mine of its assets and removes them before the authorities can lay claim to them. Some of this behaviour could be criminal and lawsuits are likely.

Second, the mining industry has been under pressure to improve its environmental responsibility over the last 50 years. Consequently, physical environmental aspects dominate the research on mine closure and the research originates largely from the industry (Bainton & Holcombe, 2018). The annual conferences on mine closure in Australia have helped to increase this body of work. Some contributors to these conferences have considered the wider range of socio-economic issues related to mine closure, but the tendency is to assume that the environmental issues include the socio-economic issues. Although this is a valid assumption, the environmental issues dominate. For many years, successful closure has depended largely on dealing successfully with the environmental concerns. It has only recently been accepted that the broader social issues, including social costs, should form part of planning for mine closure.

Third, mine closure does not always mean the physical closure of a mine; it could mean placing the mine under 'care and maintenance', which ensures that the mine can easily become operational again if the economic circumstances improve (Vivoda et al., 2019). Reasons for this kind of 'closure' are that it gives the mining company time to sell the mine or keep costs down while waiting for resource prices to bounce back. Care and maintenance limits mine activity, ends social programmes and causes uncertainty about when, or whether, the mine will reopen.

Fourth, mine closure can take the form of large mining companies selling their mines to smaller companies (Bainton & Holcombe, 2018). Small companies usually have a more cost-effective structure and operational procedures. The long-term problem is that they seldom have the financial backing to ensure appropriate mine closure. In addition, they are unlikely to pay attention to the social concerns of closure. The handover to small companies increases the likelihood of unstructured closure or abandoned mines, leaving the state responsible for closing them.

Fifth, mine closure processes can be unpredictable and slow. Sometimes decline and closure is quick, but usually it takes a long time. This makes predicting mine closure difficult for mining companies and even more difficult for communities and households. It entails a gradual decrease in the workforce over time and sub-periods of rising and declining employment.

Finally, some researchers describe mine closure as a process of economic restructuring (Digby, 2012). This way of thinking is based on the lesson of the Rust Belt states in the US losing their manufacturing capabilities. Mine decline can be seen as resulting from a decline in the primary sector where workplace skills depend largely on low-skilled jobs (despite some change globally, this remains the reality in Africa). It is more difficult in the mining sector than in the manufacturing sector to transfer skills to other economic sectors. In other words, economic restructuring is more complicated in mining than in manufacturing.

One commonality, however, among all the variations and complications of mine closure, is that mining companies, local governments, communities and households have to think about the social consequences.

3.2.2 *The social consequences of mine closure*

Although the literature on the social consequences of mine closure is relatively small, it has been growing recently. Mine closures are complex so many people are involved, all with different values, interests and objectives. In this section, I drew on a wide array of sources: Neil et al. (1992), Binns and Nel (2001, 2002), World Bank (2002), Limpitlaw (2004), Marais et al. (2005), Andrews-Speed et al. (2005), Limpitlaw and Hoadley (2006), Haslam McKenzie et al. (2009), Linde et al. (2012), Marais (2013a, 2013b), Petrova and Marinova (2013), Limpitlaw and Briel (2014), Bennett (2016), Marais and Nel (2016), Nehring and Cheng (2016), Chaloping-March (2017), Ntema et al. (2017), Ackerman et al. (2018), Bainton and Holcombe (2018), Gough et al. (2018), Louw and Marais (2018), Marais (2018), Marais et al. (2018), Owen and Kemp (2018), Marais et al. (2019), Van der Watt and Marais (2019), Vivoda et al. (2019), Marais et al. (2020), Crous et al. (2021), Sesele et al. (2021a) and Sesele et al. (2021b).

Concern about the social aspects of mining during the operational phase is not new. Terms like 'socially responsible mining', 'a social licence to operate' and 'corporate responsibility' are in common use by the mining industry. But researchers have argued that the concept of 'social aspects of mine closure' should include broader issues than 'social projects' or 'social impacts'. It should cover a range of socio-economic, political, cultural and institutional consequences, how communities and planners can manage them and how people imagine a post-mining economy.

Table 3.1 is an example I have borrowed from the literature on social aspects of mine closure. It lists indicators related to social aspects of mine closure. These should not be treated as separate concerns; rather, mining companies and communities should think about what *links* them. For example, falling house prices might reduce the rates and taxes that a town can levy, which will in turn reduce

Table 3.1 Social aspects of mine closure

Social aspect	*Indicators*
Economic	Local economic activity (diversity and dependence) Household income Local living standards
Business	Local business development opportunities in the mining sector Business opportunities in other sectors
Employment	Local employment opportunities in the mining sector Local employment opportunities in other sectors Local employment stability/volatility
Security, education and training infrastructure	Social order and safety (e.g. tensions, crime and violence) Local skills development in mining and other sectors Access to quality education and training Local transport (e.g. public buses, roads and airports) Critical infrastructure (food supply, power supply, water supply, telecommunications)
Amenities	Local aesthetic and recreational resources (e.g. heritage sites, parks and recreation areas, communal areas) Local culture, arts and sports (including facilities)
Livelihoods	Local livelihoods (e.g. access to land, food, water and shelter that affect livelihoods)
Land	Local land access, ownership and use Recognition of traditional, customary ownership
Housing and health	Local housing quality, availability and affordability Community health and wellbeing Access to quality health and social services
Environment	Environmental aspects that affect social conditions (e.g. quality of air, water, land, ecosystem)
Demography	Local population dynamics (e.g. growth/decline, migration, ageing, gender balance)
Participation	Stakeholder participation in closure planning and closure and post-closure processes (including decision-making)
Inclusion	Inclusive stakeholder engagement, including vulnerable and otherwise marginalised groups (e.g. indigenous peoples, women, ethnic minorities, disabled, elderly, young) in closure planning and closure and post-closure processes (including decision-making)
Social (general)	General socio-economic considerations Social considerations in financial assurance mechanisms

Source: Vivoda et al. (2019, p. 8).

its ability to provide social amenities. Furthermore, mine closure changes the social and demographic makeup of communities. Increased unemployment and the outmigration of men are common at the beginning of the closure process, with inevitable negative effects on the women and children who stay behind until the man finds a stable job. Moreover, mines and communities should consider cross-cutting issues like gender and children's rights.

Vivoda et al.'s table is not an exhaustive list or the only framework available. The International Council on Mining and Metals (ICMM) good practice guide to managing mine closure, for example, provides a table listing important consequences (ICMM, 2019, pp. 82–83). It identifies 18 social categories for which the question 'What social or socio-economic values and gains can be achieved? should be asked: poverty; hunger; education; gender equality; child mortality; maternal health; HIV/AIDS, malaria and other diseases; health care; water supply; employment; youth employment; employability; technology; recreation; infrastructure; indigenous affairs; culture; and enterprise.

Besides the social cost of mine closure, there are, of course, concerns about the financial cost. Mining companies are legally required to provide detailed cost estimates related to land rehabilitation. But simply complying with this legislation is a narrow response to mine closure. The industry requires much more research to understand the social costs.

The international literature points to high levels of interdependence between mining companies and local communities, partly the effect of the mine's obligation to obtain a social licence to operate. Investment in obtaining this licence represents the industry's 'front end' approach (when a mine starts), in which substantial investments are made to ensure that mining proceeds (Vivoda et al., 2019). Little attention is devoted at the 'back end' a mine closes. There are many examples of dependencies this licence creates. For example, there are cases where mining companies have managed mining towns themselves or performed various public sector functions, such as paying the salaries of teachers or clinic staff or managing their clinics. However, once a mine closes, these dependencies cease to exist, with negative consequences for the mining community.

A small amount of research has been done on normalisation – the process of transferring towns managed by the mines to democratic governance (Van der Watt and Marais, 2019). The mining industry sees normalisation as part of the process of shifting from company dominance to collaborative planning. Governments promote this shift because it supports local control. But there are pros and cons. Normalisation will encourage local control and help people to move on from the mining industries' patriarchal system. But it could put them at risk when mines transfer their responsibilities to local government and individual households. Both normalisation and ongoing town management by mining companies create dependencies. Under normalisation, communities have to depend on the local government – and themselves – instead of on the mine.

3.2.3 Policy responses to mine closure

There is a growing body of work on how the various role players react to the social aspects of mine closure. In this section, I highlight the most important points from the literature, using the following sources: World Bank (2002), Peck et al. (2005), Didier et al. (2008), Lahiri-Dutt (2012), Boschma (2017), Chaloping-March (2017), Mikovits et al. (2018), ICMM (2019), Manzanoa and Gutiérrez (2019), Atienza, Arias-Loyola and Phelp (2021a), Atienza, Lufin and Soto (2021b) and Sesele et al. (2021b).

A common response to mine decline has been to focus on economic diversification. But it remains difficult to diversify the economies of isolated mining communities. Attempts to find alternative uses for land, reskill employees, attract different economic sectors and repurpose infrastructure have met with only limited success (Atienza et al., 2021a, 2021b). Typical problems include the dominance of the mining company, the economic dependence that mining creates in the community, the generally low level of skills in mining communities and the destruction of habitat and land. Mining economies tend to centralise in a country's gateway cities. These cities benefit from the mining value chain, although the mines themselves are usually in other areas. Mining has a limited ability to serve as a springboard for the development of other economic sectors, mainly because of the lower levels of skills and the inability to create 'relatedness' with other sectors. The term 'relatedness' refers to how new economic sectors are dependent on established sectors. It generally refers to the knowledge spillovers from established sectors to new economies in the information and communications technology sectors. Mining's limited relatedness to other sectors makes it difficult to diversify the economy away from mining (Atienza et al., 2021a). The international literature sees relatedness as a central driver for local economic diversification, but mining is unlikely to provide the necessary platform. Coal mining's relatedness to energy creation is one of very few examples of local relatedness in the mining sector.

Another common response is to link closure to the mine life-cycle planning (Chaloping-March, 2017). This thinking originates from life-cycle planning, common in the mining industry since the 1970s. It emphasises that planning for decline should be done from inception. Such a process has merits and has improved closure planning in the mining industry. Mines mostly do this efficiently, but local governments and mining communities find it difficult. The problem is how to transfer this thinking from the mining industry to the public domain. New mining development places extraordinary pressure on mining towns. That local governments should, in addition to managing the economic boom, plan 20–30 years ahead for the decline may be too much to expect. Unlike the mining companies, local governments seldom have accurate information on future mining projects and consequently struggle to plan ahead. Furthermore, the inclusion of mine closure plans at the outset is an industry response and a linear response. The implications for households and communities are more complex.

For example, mine closure ends social investment programmes but creates new social consequences which continue well beyond mine closure.

Yet another common response associated with planning decline is to emphasise stakeholder engagement during mine closure. But it is difficult to ensure participation. In addition, there are concerns that women are under-represented in impact assessments at the start of mining projects. Therefore, they will be under-represented in mine closure participation processes. Considering the overall consequences of mine closure for women, such involvement is crucial (Sesele et al., 2021b).

The ICMM (2019) guidelines suggest several responses in dealing with mine closure. They emphasise the need to include mine closure in the mine's life cycle. This requires the company to develop a closure vision, objectives and principles, and an assessment of opportunities and risks. Central to success is to engage stakeholders and to collect accurate information, keep it updated and make it available to all the affected parties. Once the vision is in place, the company must create an implementation plan, documenting the possible obstacles and successes and designed to be flexible in dealing with different closure scenarios. The existence of this plan will help to demonstrate a commitment to closure. The company must prepare the workforce and the community for social transition beyond the life cycle of mining. It must determine the cost of mine closure and provide assurance on how it will deal with contingencies in the industry. An appropriate governance structure is critical in the mine closure process, to ensure that the required resources are available once decline sets in. Closure plans and governance structures require frequent evaluation and adjustment. The aim is to minimise disruption because of unplanned events and have contingency plans for temporary or sudden closures. Final closure must include the handover of the mining sites to the responsible authorities.

The ICMM's guide is useful but not beyond criticism. As I have pointed out earlier, there is a tendency to assume that mine life-cycle planning also applies to planning for mine closure in the wider community. Mine life-cycle planning assumes a narrow and linear process. In contrast, closure planning for the community must address multifarious issues, and this is not easy. There is a tendency to assume that it is possible to mitigate mine closure. In reality, this is impossible without a long-term regime of government subsidies.

I argue that a more realistic approach would be not to assume the possibility of mitigation. This should be a central question in thinking about the guidelines from the ICMM. I note that the ICMM does not pay attention to the way a community becomes dependent on a mine. It assumes that community effects will be positive while the mine is in operation and does not consider the likelihood of these effects becoming dependencies. As I have said earlier, the challenge is to plan mine closure to minimise the long-term consequences of dependencies that mining creates during its operations.

3.2.4 *Mine closure policy in South Africa*

Until 1991, no legislation governed mine closure in South Africa. The companies simply abandoned mines when they became unprofitable. The first legislation in 1991 required land rehabilitation, and subsequent legislation has tightened up the environmental and financial requirements. However, the new legislation does not apply to abandoned mines, of which there are hundreds.

As elsewhere in the world, South Africa's policy on mine closure focuses on the environmental aspects. Below I discuss the mine closure policies as reflected in the White Paper 'A Minerals and Mining Policy for South Africa' (DME, 1998), the Mineral and Petroleum Resources Development Act (MPRDA) (2002) and the Social and Labour Plan (SLP) Guidelines (DME, 2004; DMR, 2010).

The White Paper resulted in the government accepting the MPRDA, and the MPRDA requires mining companies to develop SLPs, which are a means of institutionalising collaborative planning (Van der Watt & Marais, 2021). The 1998 White Paper, the first mining policy from South Africa's post-apartheid government, highlighted four factors the mining industry had neglected under apartheid: ownership and management, labour legislation, environmental protection and health and safety in the workplace. It is noteworthy that mine closure and its related social aspects are missing from this list. Yet the development of the White Paper coincided with dwindling profits and a decline in gold mining. The White Paper did not view this as a problem as there were growth opportunities in iron ore, manganese and coal. Its view was that while the historical problems in the industry required urgent attention, global competitiveness was still more urgent. Therefore, mine decline and closure did not receive much attention.

I have five concerns with the White Paper. First, while it acknowledges that closure is a normal process, that a mine is by nature finite and that technology and mechanisation will exhaust a resource more rapidly and jobs will be lost, it notes that closure must be monitored. However, it offers no details about the purposes of monitoring, the possible approaches or who will be responsible. Second, it acknowledges that mine downscaling and closure will have local and regional consequences but does not indicate the nature and scale of these consequences or the regional knock-on effects. Third, it says national government should play a role in mitigating the consequences, but, in line with the concept of collaborative planning, expects local governments to plan for mine closure. It does not say what potential effects the planners should seek to mitigate, nor does it take into account that local government is unlikely to be able to cope, being the sphere of government with the least capacity or funds. Fourth, highlighting the problems that mine decline and closure bring to mineworkers, it focuses on consulting the workforce and providing retraining and counselling services. Considering the range of potential social consequences of mine closure, this narrow focus on the workforce seems inappropriate. Finally, it argues that elsewhere in the world, decline and closure hold opportunities for smaller mining companies, and in South Africa this is an opportunity to broaden ownership in the mining industry. It does not take into account that these smaller companies seldom have the financial ability to ensure proper closure or mitigate the social consequences.

Overall, I find the White Paper's response to mine closure and its social consequences inadequate. Rather than helping communities to avoid dependence on the mine, it reinforces this dependence. Its emphasis on supporting the communities shows little understanding that such support can create dependence. The failure to set out a comprehensive list of the potential effects of mine closure is a major shortcoming of this White Paper.

The MPRDA does not refer to mine closure outside the ambit of the physical environment but proposes two means to deal with it: a social fund and the effective implementation of SLPs, which are intended to ensure collaborative planning and arrange funding for social aspects of mining. The MPRDA assumes that the environmental aspects include the social aspects of mine closure. This legislative inattention to the social aspects of mine closure remains a serious policy void.

The mines are supposed to integrate their SLPs with the local governments' local strategic plans, called 'integrated development plans' (IDPs) in South Africa. The SLPs are responsible for dealing with the local reality while mining takes place. The guidelines for developing SLPs do refer to mine closure, but I find their guidance for dealing with the social consequences inadequate.

These guidelines say an SLP should have plans for five main components: human resource development, mine-community development, housing and living conditions, employment equity and saving jobs and managing decline and closure. Although the first three components focus on issues while the mine is active, the guide does refer to human resource processes when decline sets in and the mine might retrench the workers. The last two components are more directly associated with mine decline or closure. Specifically, component five collates the various elements of closure. The guidelines require the mine's closure plans to focus on mining and labour sending areas. This component emphasises preventing retrenchments and regenerating local economies. The guidelines propose the institutionalisation of a 'futures forum'. This, however, is largely focused on workers and does not include representatives from the wider community. The guidelines are not specific about how to deal with the social consequences of closure.

Among the shortcomings I find in this document are too little references to issues of gender, too much focus on workers rather than communities and minimal understanding of the potential socio-economic consequences of mine closure. A major problem is that an SLP is valid until the mine receives a closure certificate. Yet the consequences of mine closure will become apparent after closure. Although an SLP remains intact if the mine decides on care and maintenance, it might be difficult for the government to enforce the provisions of the SLP, as revoking a mine's licence could make the government responsible for the mine's closure.

3.2.5 Mine closure research in South Africa

As is the case globally, the mine closure literature in South Africa is small and, where it does exist, emphasises the environmental aspects. For my discussion in this section, I used the following sources: Binns and Nel (2001, 2002), Nel and Binns (2002), Marais et al. (2005), Van Tonder et al. (2009), Nzimande and Chauke (2012), Broomes (2013), Lane and Kamp (2013), Marais (2013a, 2013b), Druten and Bekker (2017), Muswaka (2017), Ackermann et al. (2018), Marais et al. (2018), Siyongwana (2018), Watson and Olalde (2019), Crous et al. (2021), Perkins et al. (2020), Cloete and Marais (2021), Marais et al. (2021), Marais and De Lange (2021) and Sesele et al. (2021b). I identified five types of study in this literature: studies of mine closure practice, place-based studies, overviews of the social consequences of decline and closure, studies of housing policy and mine decline and studies of mining companies' disclosure of mine closure.

As I noted earlier, the international experience shows that mine closure processes are slow and very few mines close formally. Putting a mine in care and maintenance or selling it to small mining companies are two common 'closure' methods used by large companies. The studies show that both are common in South Africa, the slow processes of the South African regulatory authorities being an inhibiting factor for mine closure.

Several place-based studies have explained the economic consequences of mine decline and local responses. The research highlights the considerable local implications of mine decline and closure (Binns & Nel, 2001; Marais, 2013a, 2013b). Although there are examples of local responses to decline, success in finding alternative economic drivers is limited. Except for one study of Newcastle, KwaZulu-Natal (Marais et al., 2021), case studies of successful transition from the mining sector to other economic sectors are scant. Newcastle's success is largely the result of state assistance in the early 1990s.

The overviews of social consequences mention areas losing their economic base and political influence, increased unemployment, poverty and outmigration by men. They also mention examples of social disruption such as alcohol and drug abuse and increased crime (e.g. Sesele et al., 2021b).

Not much work has been done on housing policy in mining areas. But questions about how post-apartheid policy could solve housing problems in mining areas are increasing in the literature. The studies point out that mine closure has increased informality and urban sprawl (Marais et al., 2018; Cloete & Marais, 2021). Researchers find that mining communities have experienced a rapid increase in population and informal housing structures. They observe that dismantling the compounds has not led to appropriate housing provision for mineworkers and mining communities.

The one study of company disclosure of mine closure (Crous et al., 2021) shows that disclosure is insufficient. Mines commonly report on closure and care and maintenance of operations, but there is very little reporting on the social consequences of their closure practices.

3.3 Shrinking cities

Over the past two to three decades, population has surged in cities of the Global South. At the same time, there is increasing evidence of some cities losing populations and their economies declining. This means of course that commercial and residential buildings will be standing empty or occupied by squatters (Fol & Cunningham-Sabot, 2010; Cunningham-Sabot et al., 2014; Weaver et al., 2017). A recent increase in the literature on shrinking cities is largely of two types: studies of the decline of cities in Eastern Europe (Kabisch et al., 2010; Sander, 2014; Dubeaux and Cunningham-Sabot, 2019) and the work of the Shrinking Cities International Research Network (Cunningham-Sabot et al., 2014). In this section, I draw on these sources and also Delken (2008), Mykhnenko and Turok, (2008), Hummel (2015), Stradling (2017) and Pallagst et al. (2014).

City governors and planners usually see the decline as a problem. It reduces the tax base, revenue and employment. Most cities struggle to adapt and their failure to cope contributes to the outflow of people. We need more research on shrinking cities. The term 'shrinking cities' was first used by Mabel Walker, president of the Municipal Finance Officers Association in the US, when addressing the Association in New York in 1947 (Beauregard, 2003). At that point, the urban decline was the result of the decentralisation of neighbourhoods and commercial activities moving to the periphery. Since then, evidence has emerged of overall city decline. For example, between 1960 and 2005, 75% of 150 cities in Europe showed a population decline.

Mine closure can contribute to shrinking, although most research on shrinking cities has come from non-mining examples. I believe that the shrinking cities literature I review below offers mine closure research a useful new perspective. It shows that proper management of shrinking is a viable alternative to economic diversification and could help a town to see value in decline.

3.3.1 *Definitions, characteristics and indicators*

The first definitions of shrinking focused on deindustrialisation and suburbanisation, as decline can occur at the neighbourhood or city level, and most definitions emphasise distress (Wiechmann & Pallagst, 2012; Cunningham-Sabot et al., 2014). As the research on shrinking cities evolved, the diversity of definitions increased. Consequently, there is no single definition of a shrinking city, and the most recent discussion emphasises the multifaceted nature of decline (Pallagst et al., 2014; Silva et al., 2015). Below, I identify the common attributes of these definitions drawn from Mykhnenko and Turok (2008), Fol and Cunningham-Sabot (2010), Hollander and Nemeth (2011), Stryjakiewicz (2013), Cunningham-Sabot et al. (2014), Pallagst et al. (2014), Silva et al. (2015) and Weaver et al. (2017).

A common denominator of most definitions is population decline. Population decline coincides with a decline in the economy and land use diversity.

Researchers categorise different types of shrinking according to the reasons for decline, such as changes in the economy, suburbanisation, structural change because of globalisation, environmental change and demographic change. Most definitions emphasise that shrinking is complex, non-linear and connected with growth. Definitions of shrinking describe a downward spiral effect with negative consequences for city incomes and the ability to attract investment. Common terms used to describe this downward spiral include 'decay', 'blight', 'abandonment' and 'urban crisis'. This downward spiral can result in conflict. Nevertheless, decline, like growth, needs to be understood as a natural urban process.

Several of the sources I consulted reflect on the consequences of shrinking. They note a change in the local labour market, changes in the socio-spatial characteristics of the cities, changing housing patterns and patterns of land use, and changes in the state of the city's infrastructure, the municipal finances and the local economy. In neighbourhoods standing empty, infrastructure deteriorates and income from rates and taxes is lost. Deindustrialisation is a common result of shrinking. Shrinking can have indirect as well as direct consequences. For example, the closure of a factory could mean that skilled people leave and vulnerable households replace them. That, in turn, might result in municipalities applying policies and procedure less rigorously than in the past.

Several researchers have measured shrinking and developed indicators and measures of shrinking, taking into account prevalence, persistence, severity and geographical incidence. Common indicators are population loss, change in per capita and aggregate income, housing vacancy rates and various dimensions of social capital (the ability to act collectively). However, these measures and indicators could be developed further. It should be noted that they do not automatically predict shrinking. For example, a city does not immediately shrink spatially when the population decreases. A city can continue to grow despite population decrease and economic decline. A major problem is that the data quantity and quality and the methods vary considerably across studies, making comparisons difficult.

3.3.2 *Shrinking cities and globalisation*

The phenomenon of shrinking cities is linked to globalisation. A mix of technological, production and social processes has affected cities worldwide (Suresh, 2003; Faulconbridge and Beaverstock, 2009; Castells, 2010). Pro-globalisation policies reduce trade barriers and attract foreign direct investment and the pressure to accept global standards. Many American cities that depended on manufacturing have experienced competition from various parts of the world. The opening of global markets and new production processes, bringing a rise in transnational corporations, has affected these cities negatively. Such cities try to take advantage of globalisation, going out of their way to attract investments and skills and invest in technology. Not all succeed. Some find it difficult to stay connected, others are defeated by better opposition. As a result, they lose out on new investments and fail to attract highly skilled people.

3.3.3 Planning for decline

Knowing how cities should respond to decline and shrinkage is important (Hollander & Nemeth, 2011; Stryjakiewicz, 2013; Weaver et al., 2017). The problem is that growth is generally treated as equivalent to success and decline as a sign of failure (Popper & Popper, 2010). Getting city governors and planners to value decline is like telling them to accept failure. Good planning for decline focuses on providing a better quality of life for the reduced number of people. Below, I look at ideas on good planning for and management of decline that should become the appropriate response to shrinking. I draw on Schetke and Haase (2008), Schilling and Logan (2008), LaCroix (2010), Burkholder (2012), Haase et al. (2014), Hummel (2015), Pallagst et al. (2017), Paddeu (2017), Rupprecht (2017), Weaver et al. (2017), Ghosh et al. (2019), Joo and Seo (2018), Riley et al. (2019) and Zhang et al. (2019).

Finding appropriate ways to deal with shrinking is not easy. Four possible responses to the problem are to ignore it, to counter it, to accept it or to use it as an opportunity. Ignoring it means that the authorities do not acknowledge the problem and make no plans. A common attempt to counter the shrinking is to make growth-oriented plans. The challenge is to regain investments that have been lost. Shrinking mining towns' attempts to diversify their economy by attracting alternative economic sectors have met with little success, particularly if they are isolated and have no state subsidies. Finally, shrinking requires management. It must be used as an opportunity to create a small but more appropriate living environment. Few have recognised the opportunity of engaging with a non-growth planning paradigm.

Despite the emphasis on planning for growth, planning for a smaller and better future has received some attention. One of the main advantages of appropriate planning for decline is the potential for ecological restoration and protecting land. Mixed land use is encouraged. The overall aim is to improve the quality of life for those who remain. This contrasts with strategies that try to attract new people to the area. Common terms used in this approach are 'smart decline' and 'smart shrinkage', derived from the 'smart' growth concept that has become dominant over the past two decades. For various reasons, such as different local contexts, smart decline is not always well defined, but there are several commonalities. Most shrinking cities have to rethink everything, from transport and energy systems to land use patterns, rightsizing and ecological restoration. The focus is on increasing urban density, despite the decline, and improving the quality of life. Decline requires authorities to rethink their approach to public services and infrastructure provision, closure of schools and health facilities, austerity measures, urban restructuring and finding public and private finances to manage the decline. Social justice is central to planning smart decline.

The literature points to five elements of decline planning: recognition of multiple voices, a political and deliberate process, the use of a variety of communication techniques and enabling citizens to challenge power imbalances, transparency and a regional scope but local control. Everybody involved in the process must have a realistic understanding of planning for decline. Participation from different interest groups is important and planners should be able to listen to everybody. However, smart decline has been criticised. Common criticisms are that it takes a top-down approach, fails to engage with the public and assumes that all shrinking cities are the same.

Many issues need to be considered in planning for decline. Rightsizing is important, as vacant and abandoned land is difficult to maintain. It could include demolishing or repurposing abandoned buildings. Dysfunctional markets must be stabilised. This usually goes hand in hand with promoting green infrastructure and land uses and adjusting the amount of land available for development. Overall, rightsizing should contribute to green infrastructure and programmes, create a land bank and build community consensus.

Shrinking requires that local governments change their approach to infrastructure and land use management and learn to view shrinkage in a positive light. The literature points to three ways of taking a positive approach to shrinking. First, the local planning system should try to relink people with their physical environment. Urbanisation damages the environment by removing vegetation and creating heat islands. Shrinking creates an opportunity to revegetate urban areas, encourage biodiversity, promote cooling, improve rainwater absorption and reduce the danger of flooding. Green urban environments have wellbeing benefits. Second, shrinking has the benefit of highlighting the temporality of much land use. Mining is an obvious example of temporary land use. Decline requires flexibility in land use, thinking about possible future land uses and being able to change land use quickly. Third, planning for shrinking requires adjusting the built environment to future populations. Generally, it emphasises higher densities (and smaller land use for residential purposes) and meaningful relationships between people and open spaces. In some cases, it could result in the demolition of housing stock and generally requires decentralised infrastructure systems.

Although links exist between mine closure and the shrinking cities literature, they remain underdeveloped. The mine closure literature could benefit from a closer link. A specific focus on land use temporality and flexible change could be central to this link.

3.4 Conclusion

Governments and mining companies expect mining to create stability and continuity. This explains housing policies that encourage place attachment and accumulation of assets (see Chapter 2). However, the sense of prosperity created by a mine is a delusion, because the mine will inevitably close some day. Mine closure can create forms of social disruption that have not received much attention in the research on resource-based communities. This chapter discussed the nature and implications of mine decline and closure. Although there are examples of rapid change, most mine decline is slow, with boom and bust phases. But closure eventually comes. The challenge is not to create too many dependencies while mining is in process. Although there are good reasons to use mining as a springboard to diversify the economy, the reality in many isolated mining communities is that success is limited. Consequently, planning for shrinking remains a viable option for dealing with mine closure and decline. Too often, mine decline is seen as an opportunity to diversify economies, while planners rarely consider the benefits of mine decline and closure from a shrinking city's perspective.

This chapter suggests several policy options. First, planning for decline and closure is a realistic alternative to economic diversification in mining areas. Managing decline in appropriate ways can stimulate growth at a later stage. Second, collaborative planning between the mines and local governments should ensure mining booms do not create long-term dependencies. Policies seldom consider this principle. The question is whether a plan made during a boom can minimise long-term dependencies when decline sets in. For example, how do we plan infrastructure that does not become a liability when decline sets in? How do we plan housing so that it provides flexibility for households that need it in decline periods? Most mining companies are good at managing their long-term risks, but local communities and households are less prepared and mining gives them a false sense of prosperity. Third, resource-based communities can learn much from the way mining companies include closure in the mine's life-cycle planning. They must be helped to understand the complexity of mine closure processes and the social consequences.

Finally, there is a need to expand the list of social consequences and document them. The mine closure literature needs more systematic mine closure case studies to understand the complex and interrelated effects and the planning responses.

References

Ackermann, M., Botha, D. & van der Waldt, G., 2018. Potential socio-economic consequences of mine closure. *The Journal for Transdisciplinary Research in Southern Africa*, 14(1), p. 458.

Andrews-Speed, P., Ma, G., Shao, B. & Liao, C., 2005. Economic responses to the closure of small-scale coal mines in Chongqing, China. *Resource Policy*, 30, pp. 39–54.

Atienza, M., Arias-Loyola, M. & Phelp, N., 2021a. Gateways or backdoors to development? Filtering mechanisms and territorial embeddedness in the Chilean copper GPN's urban system. *Growth and Change*, 52, pp. 88–110.

Atienza, M., Lufin, M. & Soto, J., 2021b. Mining linkages in the Chilean copper supply network and regional economic development. *Resources Policy*, 70, p. 101154.

Bainton, N. & Holcombe, S., 2018. A critical review of the social aspects of mine closure. *Resources Policy*, 59, pp. 368–478.

Beauregard, R., 2003. *Voices of Decline: The Postwar Fate of US Cities*. London: Routledge.

Bennett, K., 2016. Abandoned mines—environmental, social and economic challenges. In: A. Fourie & M. Tibbett, eds. *Proceedings of the 11th International Conference on Mine Closure*. Perth: Australian Centre for Geomechanics, pp. 241–252.

Binns, T. & Nel, E., 2001. Gold loses its shine: decline and response in the South African goldfields. *Geography*, 86, pp. 255–260.

Binns, T. & Nel, E., 2002. The village in the game park: community response to the demise of coal mining in KwaZulu-Natal, South Africa. *Economic Geography*, 79, pp. 41–66.

Boschma, M., 2017. Relatedness as driver of regional diversification: a research agenda. *Regional Studies*, 51(3), p. 351–364.
Broomes, V., 2013. *Governance, Risk and Stakeholder Engagement: What Lessons Can Be Learnt from Mining?* https://www.researchgate.net/publication/293249990.
Burkholder, S., 2012. The new ecology of vacancy: rethinking land use in shrinking cities. *Sustainability*, 4, pp. 1154–1172.
Castells, M., 2010. *The Rise of the Network Society*. West Sussex, United Kingdom: Wiley-Blackwell.
Chaloping-March, M., 2017. *Social Terrains of Mine Closure in the Philippines*. London: Routledge.
Cloete, J. & Marais, L., 2021. Mine housing in the South African coalfields: the unforeseen consequences of post-apartheid policy. *Housing Studies*. https://doi.org/10.1080/02673037.2020.1769038.
Crous, C., Owen, J., Marais, L., Khanyile, S. & Kemp, D., 2021. Public disclosure of mine closures by listed South African mining companies. *Corporate Social Responsibility and Environmental Management*, 28(3), pp. 1032–1042.
Cunningham-Sabot, E., Audirac, I., Fol, S. & Martinez-Fernandez, C., 2014. Theoretical approaches of 'Shrinking cities'. In: K. Pallagst, T. Wiechmann, & C. Martinez-Fernandez, eds. *Shrinking Cities - International Perspectives and Policy Implications*. New York: Routledge, pp. 14–32.
Delken, E., 2008. Happiness in shrinking cities in Germany. A research note. *Journal of Happiness Studies*, 9, pp. 213–218.
Didier, C., Van der Merwe, N., Betournay, M., Mainz, M., Kotyrba, A., Oydan, O., Josien, J. & Song, W., 2008. *Mine Closure and Post-mining Management. International State-of-the-art*. International Society of Rock Mechanics (ISRM) Mine Closure Commission.
Digby, C., 2012. Mine closure through the 21st century looking glass. In: A. Fourie, ed. *Proceedings of the 7th International Conference on Mine Closure*. Perth: Australian Centre for Geomechanics, pp. 33–38.
Dubeaux, S. & Cunningham Sabot, E., 2019. Maximizing the potential of vacant spaces within shrinking cities, a German approach. *Cities*, 75, pp. 6–11.
DME (Department of Minerals and Energy), 1998. *White Paper: A Minerals and Mining Policy for South Africa*. Pretoria: Government Printer.
DME (Department of Minerals and Energy), 2004. *Guidelines for the Development of Social and Labour Plans*. Pretoria: DME.
DMR (Department of Mineral Resources and Energy), 2010. *Guidelines for the Development of Social and Labour Plans* [revised version of DME, 2004]. Pretoria: DMR.
Druten, E.v. & Bekker, M., 2017. Towards an inclusive model to address unsuccessful mine closures in South Africa. *Journal of the Southern African Institute of Mining and Metallurgy*, 117(5), pp. 485–490.
Faulconbridge, J.R. & Beaverstock, J.V., 2009. Globalization: interconnected worlds. In: N. Clifford, ed. *Key Concepts in Geography*, 2nd edn. London: Sage, pp. 331–343.
Fol, S. & Cunningham-Sabot, E., 2010. Urban decline and shrinking cities: a critical assessment of approaches to urban shrinkage. *Annales de Géographie*, 674(4), pp. 359–383.
Ghosh, S., Byahut, S. & M.C., 2019. Metropolitan regional scale smart city approaches in a shrinking city in the American rust belt—case of Pittsburgh, Pennsylvania. In: V. Kumar, ed. *Smart Metropolitan Regional Development*. New York: Springer, pp. 979–1021.
Gough, K., Yankson, P. & Esson, J., 2018. Migration, housing and attachment in urban gold mining settlements. *Urban Studies*, 56(13), pp. 2670–2687.

Haase, A. Rink, D., Grossmann, K., Bernt, M. & Mykhnenko, V., 2014. Conceptualizing urban shrinkage. *Environment and Planning A*, 46(7), pp. 1519–1534.

Haslam McKenzie, F. Phillips, R., Rowley, S., Brereton, D. & Birdsall-Jones, C., 2009. *Housing Market Dynamics in Resource Boom Towns*. Perth: Australian Housing and Urban Research Institute.

Hollander, J.B. & Nemeth, J., 2011. The bounds of smart decline: a foundational theory for planning shrinking cities. *Housing Policy Debate*, 21(3), pp. 349–367.

Hummel, D., 2015. Right-sizing cities in the United States: defining its strategies. *Journal of Urban Affairs*, 37(4), pp. 397–409.

ICMM (International Council on Mining and Metals), 2019. *Integrated Mine Closure: Good Practice Guide*, 2nd edn. London: ICMM. https://www.icmm.com/en-gb/guidance/environmental-stewardship/integrated-mine-closure-2019.

Joo, Y.-M. & Seo, B., 2018. Dual policy to fight urban shrinkage: Daegu, South Korea. *Cities*, 73, pp. 128–137.

LaCroix, 2010. Urban agriculture and other green uses: remaking the shrinking city. *The Urban Lawyer*, 42(2), pp. 225–285.

Kabisch, N., Haase, D. & Haase, A., 2010. Evolving reurbanisation? Spatio-temporal dynamics as exemplified by the East German city of Leipzig. *Urban Studies*, 47(5), 967–990.

Lahiri-Dutt, K., 2012. Digging women: towards a new agenda for. *Gender, Place & Culture*, 19(2), pp. 193–212.

Lane, A. & Kamp, R., 2013. Navigating above-the-ground risk in the platinum sector. *Journal of the Southern African Institute of Mining and Metallurgy*, 113(3), 2411–9717.

Limpitlaw, D., 2004. *Mine Closure as a Framework for Sustainable Development*. Johannesburg: School of Mining Engineering, University of the Witwatersrand.

Limpitlaw, D. & Briel, A., 2014. Post-mining land use opportunities in developing countries–a review. *Journal-South African Institute of Mining and Metallurgy*, 114(11), pp. 899–903.

Limpitlaw, D. & Hoadley, M., 2006. South Africa: faultlines in mine closure planning in a developing country context. In: A. Fourie & M. Tibbett, eds. *Proceedings of the First International Conference on Mine Closure*. Perth: Australian Centre for Geomechanics: First International Seminar on Mine Closure, pp. 803–810.

Linde, S., Matti, S. & Jagers, S.C., 2012. *Political and Institutional Prerequisites for Successful Mining Establishment and Development*. Sweden: Political Science Unit, Luleå University of Technology.

Louw, H. & Marais, L., 2018. Mining and municipal finance in Gamagara, an open mining town in South Africa. *The Extractive Industries and Society*, 5(3), pp. 278–283.

Manzanoa, O. & Gutiérrez, J., 2019. The subnational resource curse: theory and evidence. *The Extractive Industries and Society*, 6, pp. 261–266.

Marais, L., 2013a. The impact of mine downscaling in the Free State Goldfields. *Urban Forum*, 24, 503–521.

Marais, L., 2013b. Resources policy and mine closure in South Africa. The case of the Free State Goldfields. *Resources Policy*, 38, 363–372.

Marais, L., 2018. Housing policy in mining towns: issues of race and risk in South Africa. *International Journal of Housing Policy*, 18(2), pp. 335–345.

Marais, L., Cloete, J., van Rooyen, D., Denoon-Stevens, S. & Nel, V., 2019. Place attachment and social disruption in Postmasburg, a rapidly growing South African mining town. *GeoJournal*, 84(1), pp. 71–83.

Marais, L. & De Lange, A., 2021. Anticipating and planning for mine closure in South Africa. *Futures*, 25, p. 102669.

Marais, L., Denoon-Stevens, S. & Cloete, J., 2020. Mining towns and urban sprawl in South Africa. *Land Use Policy*, 98, p. 103953.
Marais, L., Haslam McKenzie, F., Deacon, L., Nel, E., Van Rooyen, D. & Cloete, J., 2018. The changing nature of mining towns: reflections from Australia, Canada and South Africa. *Land Use Policy*, 76, pp. 779–788.
Marais, L. & Nel, E., 2016. The dangers of growing on gold: lessons from the history of the Free State Goldfields, South Africa. *Local Economy*, 31(1–2), pp. 282–298.
Marais, L., Nel, V., Rani, K., Van Rooyen, D., Sesele, K., Van der Watt, P. & Du Plessis, L., 2021. Economic transitions in South Africa's secondary cities: governing mine closures. *Politics and Governance*, 9(3), pp. 381–392.
Marais, L., Pelser, A., Botes, L., Redelinghuys, N. & Benseler, A., 2005. Public finances, service delivery and mine closure in Koffiefontein (Free State, South Africa): from stepping stone to a stumbling block. *Town and Regional Planning*, 48, pp. 5–16.
Mikovits, C., Rauch, W. & Kleidorfer, M., 2018. Importance of scenario analysis in urban development for urban water infrastructure planning and management. *Computers, Environment and Urban Systems*, 68, pp. 9–16.
Muswaka, L., 2017. *An Analysis of the Legislative Framework Concerning Sustainable Mining in South Africa*. Vanderbijlpark: Faculty of Human Sciences, Vaal University of Technology.
Mykhnenko, V. & Turok, I., 2008. East European cities—patterns of growth and decline, 1960–2005. *International Planning Studies*, 13(4), pp. 311–342.
Nehring, M. & Cheng, X., 2016. An investigation into the impact of mine closure and its associated cost on life of mine planning and resource recovery. *Journal of Cleaner Production*, 127, pp. 228–239.
Neil, C., Tykkylainen, M. & Bradbury, J., 1992. *Coping with Closure: An International Comparison of Mine Town Experiences*. London: Routledge.
Nel, E. & Binns, T., 2002. Decline and response in South Africa's Free State Goldfields: local economic development in Matjhabeng. *International Development Planning Review*, 24(3), pp. 249–269.
Ntema, J., Marais, L., Cloete, J. & Lenka, M., 2017. Social disruption, mine closure and housing policy: evidence from the Free State Goldfields, South Africa. *Natural Resources Forum*, 41(1), pp. 31–40.
Nzimande, Z. & Chauke, H., 2012. Sustainability through responsible environmental mining. *Journal of the Southern African Institute of Mining and Metallurgy*, 12, pp. 135–139.
Owen, J. & Kemp, D., 2018. *Mine Closure and Social Performance: An Industry Discussion Paper*. Brisbane: Centre for Social Responsibility in Mining, Sustainable Minerals Institute, The University of Queensland.
Paddeu, F., 2017. Legalising urban agriculture in Detroit: a contested way of planning for decline. *Town Planning Review*, 88, pp. 109–129.
Pallagst, K., Fleschurz, R. & Trapp, F., 2017. Greening the shrinking city—policies and planning approaches in the USA with the example of Flint, Michigan. *Landscape Research*, 42(7), pp. 716–727.
Pallagst, K., Martinez-Frenandez, C. & Wiechmann, T., 2014. Introduction. In: K. Pallagst, T. Wiechmann, & C. Martinez-Fernandez, eds. *Shrinking Cities: International Perspectives and Policy Implications*. New York: Routledge, pp. 3–14.
Peck, P., Balkau, F., Bogdanovic, J., Sevaldsen, P., Skaalvik, J., Simonett O., Thorsen, T., Kadyrzhanova, I. & Svedberg, P., 2005. *Mining for Closure: Policies and Guidelines for Sustainable Mining Practice and Closure of Mines*. UNEP, UNDP, OSCE & NATO.

Perkins, D., Cooper, T., Scholtz, L. & Mulaudzi, K., 2020. *Mine Closures and Rehabilitation in South Africa: Activating Coalitions for the Willing.* Cape Town: WWF – World Wide Fund for Nature.
Petrova, S. & Marinova, D., 2013. Social impacts of mining: changes within the local social landscape. *Rural Society*, 22(2), pp. 153–165.
Popper, D. & Popper, F., 2010. Smart decline in post-carbon cities. Santa Rosa, CA: Post Carbon Institute.
Riley, C., Perry, K., Ard, K. & Gardiner, M., 2019. Asset or liability? Ecological and sociological tradeoffs of urban spontaneous vegetation on vacant land in shrinking cities. *Sustainability*, 10(7), pp. 2139.
Rupprecht, C., 2017. Informal urban green space: residents' perception, use, and management preferences across four major Japanese shrinking cities. *Land*, 6, pp. 3–59.
Sander, N., 2014. Internal migration in Germany, 1995–2010: new insights into East-West migration and re-urbanisation. *Comparative Population Studies*, 39(2), 217–246.
Schetke, S. & Haase, D., 2008. Multi-criteria assessment of socio-environmental aspects in shrinking cities. Experiences from eastern Germany. *Environmental Impact Assessment Review*, 28, pp. 483–503.
Schilling, J. & Logan, J., 2008. Greening the rust belt. A green infrastructure model for rightsizing America's shrinking cities. *Journal of the American Planning Association*, 74(4), pp. 451–466.
Sesele, K., Marais, L. & van Rooyen, D., 2021a. Women and mine closure: a case study of policy in South Africa. *Resources Policy*, 72, p. 102059.
Sesele, K., Marais, L., van Rooyen, D. & Cloete, J., 2021b. Mine decline and women: reflections from the Free State Goldfields. *The Extractive Industries and Society*, 8(1), pp. 211–219.
Silva, E., Pinho, P. & Reis, J.P., 2015. Measuring space: a review of spatial metrics for urban growth and shrinkage. In: M. Gunder, ed. *The Routledge Handbook of Planning Research Methods*. London: Routledge.
Siyongwana, P., 2018. The socio-economic impacts of mine closure on local communities: evidence from Mpumalanga Province in South Africa. *GeoJournal*, 84, pp. 367–380.
Stacey, J., Naude, A., Hermanus, M. & Frankel, P., 2010. The socio-economic aspects of mine closure and sustainable development: literature overview and lessons for the socio-economic aspects of closure—report 1. *The Journal of The Southern African Institute of Mining and Metallurgy*, 110, pp. 379–394.
Stradling, D., 2017. Chapter 5: The new Cuyahoga: straightening Cleveland's crooked river. In: M. Knoll, U. Lubken, & D. Schott, eds. *Rivers Lost, Rivers Regained: Rethinking City-river Relations*. University of Pittsburgh Press, pp. 107–122.
Stryjakiewicz, T., 2013. The process of urban shrinkage and its consequences. *Romanian Journal of Regional Sciences*, 7, pp. 29–40.
Suresh, B.S. Globalization and urban environmental issues and challenges. In: M. J. Bunch, V. M. Suresh, & T. Vasantha Kumaran, eds., *Proceedings of the Third International Conference on Environment and Health*, 15–17 December, 2003. Chennai: Department of Geography, University of Madras and Faculty of Environmental Studies, York University, pp. 557–561.
Van der Watt, P. & Marais, L., 2019. Normalising mining company towns in Emalahleni, South Africa. *The Extractive Industries and Society*, 6, pp. 1205–1214.
Van der Watt, P. & Marais, L., 2021. Implementing social and labour plans in South Africa: reflections on collaborative planning in the mining industry. *Resources Policy*, 71, p. 101984.

Van Tonder, D. Coetzee, H., Esterhuyse, S., Strachan, L., Wade, P. & Mudau, S., 2009. South Africa's challenges pertaining to mine closure—development and implementation of regional mining and closure strategies. In: A. Fourie & M. Tibbett, eds. *Proceedings of the 11th International Conference on Mine Closure*. Perth: Australian Centre for Geomechanics pp. 79–92.

Vivoda, V., Kemp, D. & Owen, J., 2019. Regulating the social aspects of mine closure in three Australian states. *Journal of Energy & Natural Resources Law*, 37(4), pp. 405–424.

Wang, L., Awuah-Offei, K., Que, S. & Yang, W., 2016. Eliciting drivers of community perceptions of mining projects through effective community engagement. *Sustainability*, 8(658), pp. 1–17.

Watson, I. & Olalde, M., 2019. The state of mine closure in South Africa – what the numbers say. *Journal of the Southern African Institute of Mining and Metallurgy*, 117, pp. 639–645.

Weaver, R., Bagchi-Sen, S., Knight, J. & Frazier, A.E., 2017. *Shrinking Cities – Understanding Urban Decline in the United States*. New York: Routledge.

Wiechmann, T. & Pallagst, K.M., 2012. Urban shrinkage in Germany and the USA: a comparison of transformation patterns and local strategies. *International Journal of Urban and Regional Research*, 36(2), pp. 261–280.

World Bank, 2002. *It's Not Over When It's Over: Mine Closure around the World*. Washington, DC: World Bank Group's Mining Department, Global Mining.

Zhang, Z., Meerow, S., Newell, J. & Lindquist, M., 2019. Enhancing landscape connectivity through multifunctional green infrastructure corridor modelling and design. *Urban Forestry and Urban Greening*, 83, pp. 605–511.

4 Post-apartheid housing policy and mining towns

4.1 Introduction

As I explained in Chapter 1, we can easily become stuck with the consequences of yesterday's decisions, the way we are stuck with the unchallengeable QWERTY keyboard. The post-apartheid government's decision to encourage homeownership in mining towns will have some unforeseen long-term consequences. Place attachment, which owning a home naturally encourages, is generally considered a good thing, but the dependency it creates may not necessarily be good. Getting 'locked into' a town whose continuing economic viability is doubtful can have negative consequences.

As explained in Chapter 2, homeownership and length of stay are two crucial contributors to place attachment. Accumulating assets is another. South African housing policy is an asset-based policy, helping poor people to own their homes and use their property to climb the housing ladder. I do not quarrel with the concept of asset-based development. But I challenge the assumption that government policy should blindly apply the concept to mining settlements. Housing policy for mining areas should be more nuanced. Promoting place attachment and housing asset generation could create dependencies that will become a burden in the volatile mining context. Families could find themselves locked into a town with little economic potential, and the lower skilled are likely to be the most badly affected.

I start the chapter with an analysis of mine housing under apartheid and colonial rule, to lay a foundation for understanding the post-apartheid government's responses. I then discuss the way post-apartheid housing policy in South Africa has developed, particularly in the mining context. In the final section of the chapter, I review the mine housing literature.

I make four main arguments in this chapter. First, the blind application of post-apartheid housing policy and the new mining settlement policy to mining areas puts households and local governments at risk. Government policy is to use housing to stabilise mining communities and boost local economies. Mining companies and unions have gladly accepted this policy. However, neither government, companies nor the unions consider the likelihood of mine downscaling or closure and how the current policies would affect a mining town's ability to manage the inevitable problems. The policies have relieved the mining companies of

DOI: 10.4324/9781003262015-4

the responsibility for housing and servicing mining communities, without considering that the benefits to the community also carry a substantial amount of risk (Owen et al., 2021). Second, as the post-apartheid policy was the explicit opposite of apartheid policy with the goal of making reparations, it inevitably creates goal dependency. In aspiring to undo the adverse effects of previous policies, the government fails to recognise the complexities of housing or to take into account the effects of mine closure in a mining area. Third, the new policy fails to recognise that migrant labour continues. In this chapter, I use migrant labour as one example of path dependency while at the same time arguing that the goal dependency in policy is likely to create new path dependencies. Fourth, the policy shift that has given mineworkers a choice of housing has changed the interdependencies in mine housing. Historically, the government regulated mine housing. Most mineworkers lived either in mine hostels or mine-provided housing. The mine and the government decided where mineworkers could live. Giving mineworkers a choice has rapidly increased the number of decision-makers. Every mineworker is now a decision-maker and this has eroded mine and government control.

4.2 Mine housing under colonial and apartheid rule

A fair quantity of work is available on mine housing policies under colonial and apartheid rule (e.g. Crush, 1989a, 1989b; Laburn-Peart, 1990; Crush & James, 1991; Mabin, 1991; Crush, 1992, 1994). I do not attempt to repeat everything here but simply highlight three main characteristics and the changes since the mid-1980s.

First, the system discriminated against black people. Understanding the historical racial intent lays the foundation for understanding the post-apartheid policy response. Black mineworkers were housed in single-sex hostels. Up to the mid-1980s, family housing provided by the mines was available only to white employees and company towns were primarily for whites. Only white workers had a living-out allowance and could opt for company housing or provide their own. The supervisors at the mines were generally white, and whites employees received higher salaries than their black counterparts (Nkosi, 2017). The post-apartheid policy attempt to correct the imbalances is therefore understandable.

Second, the system depended on migrant labour. Influx control prevented black people from settling in the towns. Urban jobs gave black workers permission to be in the towns but enforced back and forth migration as they had to keep their rural links. The migrant labour originated from inside South Africa (particularly from the Eastern Cape and Kwazulu-Natal) and southern Africa (Lesotho, Mozambique and Malawi). Research has documented the social cost of migrant labour in South Africa (e.g. Wilson, 1972; Murray, 1980) and I do not intend to repeat those findings here. Post-apartheid policy deliberately countered the migrant labour system by providing homeownership and integrating mining and non-mining communities into open towns.

Third, the single-sex hostel, the most common housing type for black mineworkers, prevented black urbanisation and controlled the black workforce. It reinforced the temporary nature of urban work for blacks. It saved the government the responsibility of managing influx into towns near the mines and left this task to the mining companies. The structure and layout of the hostels aided control. Living conditions were poor, with from six to 12 men per room. The conditions contributed to poor health, with TB and HIV/AIDS being common (Stuckler et al., 2013). Dismantling the single-sex hostel was thus central to post-apartheid policies.

The mid-1980s brought important changes. The apartheid government repealed influx control and reintroduced urban homeownership for black people, though only in the black townships. It was not until 1991 that the Group Areas Act was repealed, opening all areas to homeownership for blacks. From 1994, hostel dwelling was replaced mainly by homeownership and the open town policy, making urban family housing available for mineworkers.

In criticising the inhumane policies described above, researchers tend to overlook some important things about the hostels and migrant labour. They assume that temporality is necessarily bad. They fail to consider that it might have value. Under apartheid, temporality meant not being able to urbanise formally. But this temporality could be beneficial in a mining town. It could aid relocation if a mine closes and minimise social disruption. They fail to consider that the dismantling of the hostels, giving thousands of mineworkers the option to make their own housing decisions, meant that the previously simple matter of housing mineworkers became complex overnight. And they fail to consider that many mineworkers still prefer migrant work and have little intention of settling permanently near the mines.

4.3 Post-apartheid housing policy

There is an extensive body of work on post-apartheid housing policy in South Africa (Tomlinson, 1997, 1998, 1999; Huchzermeyer, 2001a, 2001b, 2001c; Tomlinson, 2002; Charlton, 2004; Charlton & Kihato, 2006; Huchzermeyer & Karam, 2006; Tomlinson, 2006, 2007; Charlton, 2009; Huchzermeyer, 2009, 2011; Charlton, 2013; Marais & Ntema, 2013; Marais & Cloete, 2015; Venter et al., 2015). Yet very little of it refers to mine housing. The government has been attempting to develop a new White Paper for Human Settlements, but at the time of writing, it had not yet seen the light of day and the current White Paper was the 1994 New Housing Policy and Strategy (Department of Housing, 1994). In 2004, the government released a revised policy approach called 'Breaking New Ground: A comprehensive plan for the development of sustainable human settlements' (commonly referred to as 'BNG') (Department of Housing, 2004). Here, I do not discuss these policy documents and the changes in detail, but rather focus on four policy aims that have relevance for mining towns: housing for social stability, homeownership, housing for asset building and private sector finance.

Both policy documents view housing as a way to create stability in society. South African policy essentially assumes an asset-based approach. The White

Paper notes that housing policy, through the quantity and quality of housing delivered, affects 'socio-political stability, productivity and attitudes and behaviour' and that housing backlogs contribute to 'the high levels of criminality and instability' (Department of Housing, 1994, pp. 5, 11). Stability was a prerequisite for banks to provide mortgages in former black township areas. BNG argues that this stability can be achieved only by 'increasing control in the hands of individuals' (Department of Housing, 2004, p. 13). BNG also uses the term 'stabilisation': it calls for the stabilisation of informal settlements and their integration 'into the broader urban fabric' (Department of Housing, 2004, p. 17). Post-apartheid policy has thus entrenched the notion that social stability can be attained through housing.

There is an emphasis on homeownership in both the White Paper and BNG. How homeownership became a dominant aspect of South African housing policy is well documented. Under apartheid, urban homeownership was available only to white people. Black urban residents could only rent. When homeownership became available for black people in the mid-1980s, it replaced rental housing as the primary housing tenure approach. This trend continued in post-apartheid policies. The result was that the largest share of housing funds favoured homeownership programmes, although rental programmes were still available. This focus on homeownership is understandable, considering the role that rental housing played in denying black people access to urban areas in South Africa. The severe problem of non-payment of monthly rentals was another reason why the government turned to homeownership in the mid-1980s as a more feasible approach.

The principle of using housing for asset-building or creating property had already appeared in the White Paper, and BNG re-emphasised it, referring to people 'climbing the housing ladder'. One of BNG's objectives is to ensure that property 'can be accessed by all as an asset for wealth creation and empowerment' (Department of Housing, 2004, p. 7). This focus on property assumes that there will be long-term economic prosperity in a mining area and that a mineworker can buy a house and sell it when leaving the area – in other words, that the house will increase, or at least retain, its value.

The housing policy also emphasises the importance of private sector finance. BNG states explicitly that 'Housing finance from financial institutions is key to the development of the primary and secondary housing market' (Department of Housing, 2004, p. 9). Getting housing finance to the lower-middle class and former black township areas has a long and turbulent history in South Africa. Much progress has been made in the past decade to stabilise housing finance in lower-income suburbs. Mine closure could derail this progress, as it has done in Thabong, the former black township of Welkom.

The White Paper lists four instruments to be used to achieve those four policy aims: project subsidies, consolidation subsidies, individual subsidies and institutional subsidies. The first three are dependent on homeownership, while the fourth is to promote rental housing. BNG makes small changes to these four instruments – the Integrated Residential Development Programme replaces project subsidies and finance-linked individual subsidies replace individual subsidies – but the emphasis on homeownership remains. At the time of writing, I was not aware of any institutional subsidies used at scale in mining areas (see also Cloete et al., 2009). This lack of rental housing provision in mining areas means that the homeownership programmes dominate housing provision in those areas.

Considering the history and the policy response after apartheid, the question is what the new policy implies for mining areas. I do not make a normative judgement against or in favour of the policy's principles; rather, I argue that applying these principles blindly in mining areas could be problematic. Using housing to create stability is not new to housing policy in South Africa or internationally. Many governments have taken this route. Against the background of South Africa's turbulent history and the instability in its mining areas, stabilisation is a rational policy response. Chapter 2 cites literature providing empirical evidence that homeownership creates place attachment and stability. Stability also helps in ensuring housing finance and a functional housing market. However, mine closure could disrupt this stability. Of course, other sectors also have disrupters that affect housing and housing finance. An important difference is that in many mining towns, mining is the dominant sector. If the mine goes, its redundant employees will struggle to pay their mortgages and maintain or renovate their property. People in other economic sectors dependent on mining could suffer the same fate. The much vaunted stability that homeownership is supposed to bring could translate to rigidity, locking owners into a declining town and causing social instability.

South Africa has seen evidence of how that can happen. The tumbling of the housing market in the Free State Goldfields from the mid-1990s is the best example (Marais, 2013; Marais & Nel, 2016; Marais et al., 2017). By 2010, house prices here were about 50% of prices for comparable houses in towns of a similar size. Two factors contributed to the fall of the housing market: the rapid economic decline because of the closure of gold mines, and mining companies dumping large numbers of mine-owned houses onto the market in the early 1990s. Supply hugely outweighed demand. The housing market decline affected households that were both directly and indirectly dependent on mining. The banks redlined Thabong, the former black township near Welkom, in the early 1990s after the first large-scale retrenchments in the area. Ntema et al. (2017) argue that homeownership locked low-income people into locations with limited economic potential at the lower end of the Free State Goldfields market. Admittedly, the relatively isolated location of the Free State Goldfields has been a contributing reason. Marais et al. (2017) have shown that mine closure in the Free State Goldfields was similar in nature and scale to mine closure in the City of Matlosana. But the economic consequences were less harsh for the City of Matlosana because of its proximity to Gauteng and Rustenburg and its historical regional services function.

4.4 How post-apartheid housing policy is implemented in mining settlements

To discover how the post-apartheid government applies the above ideas on post-apartheid housing policy to mining settlements, I examined several policy documents: the White Paper on Minerals and Mining (1998) and the Mineral and Petroleum Resources Development Act 28 of 2002 (the MPRDA), various mining charters, the guidelines for the development of Social and Labour Plans (SLPs) (2004, 2010a), the Revitalisation of Distressed Mining Towns Programme (2015) and the housing and living conditions standard for the minerals industry (2019).

4.4.1 The White Paper on Mining and Minerals and the MPRDA

The White Paper on 'A Minerals and Mining Policy for South Africa' (DME, 1998) has six main themes, one of which is 'People Issues'. This theme covers health and safety, housing needs, migrant labour, industrial relations and downscaling. The discussion that follows focuses on housing needs and migrant labour.

I highlight six points from the White Paper that are relevant to mine housing. First, the White Paper quite rightly criticises the hostels, mentioning the poor living conditions, the racial discrimination, the level of control and the restriction to men only. It notes that workers 'were often forced to live in austere, regimented single-sex hostels, subject to strict legal and extra-legal controls' (DME, 1998, p. 45). Consequently, it proposes that government should enforce norms and standards for mine housing. It refers explicitly to the number of people per room. Second, it refers to the effect of the poor quality housing on mineworkers' health and productivity. It argues that, as better housing should improve health and productivity, it is in everybody's interest to improve the living conditions. This suggests an economic incentive for mining companies to consider upgrading their workers' housing. Third, it acknowledges that mining companies have historically invested large sums of money in urban infrastructure (although mainly for their white workforce) and that these investments were mainly in isolated locations and did have value. Fourth, it sees the value of a balance between mine-provided housing and private rental or homeownership. It says there is 'merit both in continuing to provide accommodation (and to upgrade accommodation), within the constraints of costs, and in encouraging and facilitating homeownership/rental by employees within nearby communities' (DME, 1998, p. 45). This dual provision of housing is important, as later policy and mine responses have often overemphasised homeownership models. Fifth, it is critical of company towns. It says the 'whole structure of mining towns and settlements must be altered to integrate mineworkers into the local economy and to end the racially discriminatory provisions that apply to housing for black mineworkers' (DME, 1998, p. 45). Given that, historically, company towns were primarily reserved for the white workforce, the ideal of integration is understandable. However, this represents the first sign of thinking about mineworker housing as a more permanent endeavour. Housing had to be linked to the local economy. Sixth, despite a shift to homeownership, the White Paper proposes the importance of choice and a range of tenure types: rental, homeownership and government-supported rental housing (commonly referred to as 'social housing' in South Africa). Housing typologies should include housing for single people and families and housing in open towns. As opposed to the strict control imposed under apartheid and colonial rule, the White Paper says its intent is 'to ensure that all employees have a choice in their pursuit of suitable housing and living conditions' (DME, 1998, p. 45). As noted earlier, the essence of the policy is effectively to provide choice to thousands of mineworkers.

The MPRDA makes only one reference to housing: it states that the department must develop a housing and living conditions standard for the minerals industry (see Section 4.4.5 below).

The White Paper acknowledges the long history of migrant labour in the mining industry, its social consequences, its continuation today and the consequences for labour-sending communities. Despite this acknowledgement, however, the White Paper's intention is clear: the government will phase out migrant labour. It says two practical ways the government can do this are to provide family housing and recruit more labour locally. The reality of continued migrant labour is not thoroughly considered.

Overall, the housing response in the White Paper and MPRDA is sober. Three new directions in these two policy documents – providing mineworkers with a choice, encouraging homeownership and developing open towns instead of company towns – are changing mine housing and creating new dependencies. Moving from a system of control under apartheid to giving mineworkers a choice is placing immense pressure on mining towns (Cloete & Marais, 2021). The shift has resulted in large-scale informal settlements and an increase in backyard dwellings. It has transferred the long-term responsibility and risks from mining companies to individual households and local government (Van der Watt & Marais, 2021). Section 4.5 reviews a selection of the literature on the unintended consequences of the change.

4.4.2 The mining charters

The MPRDA requires the development of a mining charter to direct and monitor the mining sector's transformation. The Department of Minerals and Energy (DME) released the first mining charter in 2002 and it was amended by the Department of Mineral Resources (DMR) in 2010 (DME, 2002; DMR, 2010b). A new version was released in 2017 and finalised in 2018 (DMR, 2018).

In line with the 1998 White Paper, the mining charters aimed to improve the mineworkers' living conditions. Despite the White Paper's original recommendation of a variety of types of tenure, the first charter focused mainly on family housing and homeownership, emphasising the dismantling of shared accommodation in single-sex hostels, and specifically on developing mining communities by providing mineworkers with a choice. The provision of living-out allowances (a monthly allowance for all mineworkers who do not access mine-provided housing) is central to the 2010 and 2018 charters (Rubin & Harrison, 2016). The 2018 housing charter, the latest at the time of writing, sets out the following principles: decent and affordable housing; provision for homeownership; provision for social, physical and economic integration of human settlements; and secure tenure for mine employees in housing institutions.

Although the various charters show some toning down of the insistence on homeownership, it nevertheless remains prominent. The emphasis on developing integrated human settlements has increased, in line with other policy directions.

4.4.3 *Guidelines for Social and Labour Plans*

To integrate mining and non-mining communities, it has been necessary to develop open towns or normalise company towns, and this requires collaborative planning between the mines, the local government and the community. To ensure this planning would happen, the MPRDA introduced SLPs. The mining companies are required to produce these plans, and they have to be dovetailed with municipalities' local strategic plans called Integrated Development Plans (IDPs). The idea of IDPs is based on the New Public Management paradigm that became prominent in the 1990s. It assumes a strategic and business-like approach to planning. IDPs focus mainly on planning or 'steering', with less focus on implementation or 'rowing' (Marais et al., 2021a). This can mean that a policy's practical consequences may not be as thoroughly considered as one might wish and implementation could be neglected.

Since the promulgation of the MPRDA, the government has developed sets of guidelines for SLPs (DME, 2004; DMR, 2010a). The 2004 guidelines state that mining should avoid the development of ghost towns. The 2010 guidelines require specific targets for the promotion of homeownership and converting hostels into single quarters. There are also specific requirements for alignment with the municipality's housing plan. Economic diversification and integrating mining into the local economy, mostly in line with the 1998 White Paper, are central themes. Economic diversification requires economic planning beyond the life of a mine. The problem here is not the principle itself but how to apply it: in the isolated locations of many mines, this is no easy task. It is unlikely that many mining communities will diversify their economies. Chapter 3 has emphasised the importance of planning for decline and shrinking.

4.4.4 *The revitalisation of distressed mining towns*

The Marikana massacre of 2012, in which 34 miners were killed by the South African police, spotlighted the vulnerabilities and social instability of mining communities. The mineworkers were striking about salaries and their dire living conditions. One of the government's responses was to release the Revitalisation of Distressed Mining Towns Programme in 2015. Details of the programme are not publically available, but two documents helped me to analyse the policy direction in this programme (National Council of Provinces, 2015; Tshangana, 2015).

The programme focuses primarily on human settlement development. For example, the minutes of a 2 June 2015 meeting record that the 'overall objective was to transform mining towns through the creation of sustainable human settlements' (National Council of Provinces, 2015, p. 1). This focus is not surprising, given the proliferation of informal settlements in mining towns and the poor living conditions for mineworkers. The programme identified more than 200 informal settlements in mining areas. This focus on human settlements is also evident in the prominent allocation of housing funding to mining communities. To a large degree, the programme's strategic direction originates from the 1998 White Paper and the ideas already expressed in the guidelines for SLPs. However, unlike other policy documents which offer only tentative guidelines, this programme is far more explicit about the overall goal of housing and homeownership to bring stability (see Chapter 2). The government established an inter-ministerial task team, with the Department of Human Settlements dominating the team's management and funding. Geographically, the programme focused on 15 mining areas in five provinces (the Free State, Gauteng, Limpopo, Mpumalanga and the North West). The programme also included these areas' labour-sending areas. Mining companies provided substantial funding support to ensure the housing focus in the programme.

I have several concerns about the programme. Its developers do not appear to have a nuanced understanding of what 'distress' means in the mining context. They appear to view distress caused by mining decline in the same light as distress caused by mining growth. For example, both Matjhabeng, which is suffering mining decline, and Rustenburg, where mining is growing, are supposed to benefit from the programme and in the same way. In practice, the benefit in both cases is more and better housing and homeownership. The developers do not appear to understand that housing and land development, which the programme favours, are the *reason* for the current distress in some areas. Marais and Nel (2016) note that the 'garden city' planning ideas and the historical low-density housing developments are the main reasons for Matjhabeng's economic distress. Pushing for more housing and land development is likely to exacerbate the hardship rather than relieve it. Most of the housing projects under this programme have been homeownership projects. The programme developers assume that all mineworkers need homeownership and will take it up. They do not consider the likelihood that some of them are still migrant labourers and prefer to keep moving rather than settle. The drive to create stability is not without merit, but the developers fail to understand that with stability comes dependence. The relationship between housing, homeownership and stability lies at the heart of the problem. The rigid application of these methods to create stability, without considering how the effects may differ in different contexts, or considering other methods, is an example of the kind of path dependency that South Africa should be avoiding.

4.4.5 *Housing and living conditions standard for the mining industry, 2019*

The government released the draft housing living conditions standard for the mining industry in 2012 and then a revised version in 2019 (DMR, 2019). Here, I refer only to the latter. The MPRDA requires that government develop these standards. The mission stated in this policy document is 'to redress the historic and systematic marginalisation of mine employees' and 'to provide for progressive realisation and protection of their basic constitutional right to human dignity through provision of adequate access to housing' (DMR, 2019, p. 8). The need for appropriate housing coincides with a focus on health care and nutrition. The document sets out five principles applicable to mine housing: developing socially, physically and economically integrated housing within a mining community; creating measures to address the housing demand; involving employees in the housing administration systems; promoting best practice and compliance with minimum norms and standards for delivering and managing housing; and ensuring that financing schemes are used in a transparent and accountable manner.

Here, I summarise the principles and objectives in this document and highlight some concerns. The notion of collaborative planning is made central to planning for housing. The housing and living conditions set out in the mining companies' SLPs are to be aligned with those in the municipalities' IDPs. Furthermore, there is an expectation that government subsidies should be used to ensure the standards are achieved. Mineworkers' houses should be 'within proximity of the mining operations', but the document does not say what distance 'within proximity' means. It stresses that mineworkers should be housed in a non-racial way, which is understandable considering the history of mine housing in South Africa. It says a range of housing options should form part of the housing plan. It refers to rental housing, homeownership, government-led social housing and a living-out allowance for those who do not make use of mine housing. It says housing could be single or family housing – it frames rental housing in the context of labour mobility in the mining sector (i.e. it recognises the persistence of migrant labour). Living-out allowances should be phased out if they are not contributing to housing that meets the norms and standards and the mines should prove that the living-out allowance provides for housing that falls within the norms and standards. It says basic services must be available at the mine-provided housing – it specifically mentions water and electricity. It notes that although there has been a shift towards a 'clean wage', housing issues have not always been cleared from the wage bill (e.g. living out allowances remain in the industry). It says housing 'as part of the remuneration package should be negotiated through collective bargaining' (DMR, 2019, p. 10).

The document further emphasises the integration of mining and non-mining communities in urban settlements. This notion has been central to government policy since the release of the 1998 White Paper. It does not mention the extraordinary pressure that normalisation and open town development policies have placed on local municipalities (Louw & Marais, 2018) and on housing in mining areas (Van der Watt & Marais, 2019).

In its favour, this document does contribute to the policy debate, but it has limitations. I would expect housing and living conditions plans to reflect on mine closure and post-mining developments, but the phrase 'mine closure' occurs only once in the document. It says the housing and living conditions plans in the SLPs should reflect on mine closure and post-mining developments but offers no guidelines on how to include these concerns.

The document contradicts itself in emphasising the principle of racial integration and integrated settlements but then requiring that mineworkers' housing be situated near the mines. The focus on collaboration between the local government and the mines in their housing and living conditions plans (IDPs and SLPs) is a positive aspect. Yet, it has been shown how difficult it is in practice to build collaboration in the local sphere (Van der Watt & Marais, 2021). The linking of rental housing to labour mobility is an important contribution – this is one of the few policy documents to recognise the reality of continued migrant labour. The document also shows some understanding of the complications associated with living-out allowances. By increasing the demand for housing, living-out allowances have put pressure on mining towns (Marais & Venter, 2006; Cloete & Marais, 2021; Marais et al., 2018). Future policy debates must acknowledge that this policy direction has not necessarily improved mineworkers' housing conditions. Finally, the question must be asked whether these guidelines can really help to improve these conditions. An emphasis on norms and standards for housing has seldom helped address housing challenges. We are seeing evidence of both improvement and deterioration of housing conditions. I discuss these contradictory findings in Section 4.5 and later in the book.

4.5 The literature on mineworker housing after apartheid

Considering how much has been written on housing and informal settlement upgrading in South Africa, it is surprising to find so little on housing conditions in mining towns. Under apartheid and colonial rule, much research was done on mine housing and welfare, but post-apartheid research on these topics is sparse. Macmillan (2012, p. 539) notes that 'very little attention in the literature has been paid to the decline of gold mining in South Africa'.

To provide this overview of mineworker housing and policy debates since the demise of apartheid, I used the following works: Demissie (1998), Marais and Venter (2006), Drewes and Van Aswegen (2008), Cloete (2009), Cloete et al. (2009), Cloete and Marais (2009), Bezuidenhout and Buhlungu (2011), Macmillan (2012), Marais and Cloete (2013), Marais and Nel (2016), Rubin and Harrison (2016), Ntema et al. (2017), Denoon-Stevens and Cloete (2018), Burger and Geldenhuys (2018), Manenzhe (2018), Stewart and Drewes (2018), Marais (2018); Marais et al. (2018); Pelders and Nelson (2019), Daw (2020), Marais et al. (2020), Cloete and Marais (2021), Cole and Broadhurst (2021), Marais et al. (2021b) and Marais et al. (2022). I highlight six main issues raised in these works: housing conditions remain poor, municipalities are not coping with the influx of house-seekers, social stratification is increasing, responsibility for housing has been shifted from the mines to individuals and municipalities, migrant labour continues and policy-makers pay little attention to the threat of mine decline and closure.

There is ample evidence that housing conditions for many mineworkers remain inadequate. Receiving a living-out allowance and a choice of housing type has not necessarily improved mineworkers' housing conditions. Mineworker housing is often informal, with inadequate municipal services. Informal housing persists for several reasons. The dismantling of the single-sex hostels and closing of company town obliged mineworkers to find alternative housing. Many municipalities have been unable to manage the consequent influx of house-seekers. The proliferation of informal settlements has been an unintended consequence of the well-meaning housing policy.

The evidence points to an increase in social stratification. The difference between housing conditions for permanently employed mineworkers and contract workers is increasing. In Rustenburg, contract workers are more likely to remain in the former hostels and also more likely to live in informal housing. Differences are also observed between mineworkers and non-mineworkers in the mining towns: in most cases the former are better housed. Some of this social stratification has been the result of differentiated incomes for mineworkers. But in Postmasburg, it is also the result of mining companies investing heavily in housing for mineworkers – either rental housing or subsidised ownership.

Responsibility for mineworker housing is being shifted from the government and mining companies to the individual and municipalities. The post-apartheid policy shift from hostel accommodation, company towns and extensive company housing to housing choice for black mineworkers and bargaining for living-out allowance by unions suited the mining companies because they wanted to minimise their long-term non-mining liabilities

The research shows that a substantial portion of mineworkers are continuing with migrant labour. The scale and nature of migrant labour varies between locations. However, it is one of the contributing reasons for informal housing and a higher demand for rental housing in some mining towns.

The literature is beginning to question the overemphasis on homeownership and to analyse it in terms of goal dependency and path dependency. The vital question is: how will possible mine closure affect the focus on homeownership? Evidence from the Free State Goldfields points to the likelihood of lock-ins because of this emphasis.

4.6 Conclusion

The historical housing policy that limited black urbanisation by housing workers in hostels and enforcing migrant labour had adverse social and economic implications for black people. Overall, it prevented social stability and stopped black people benefiting from the urban premium. Addressing these two factors was central to post-apartheid policy.

Post-apartheid housing policy has three intentions: undoing the unjust historical housing for black mineworkers, applying the post-apartheid housing policy focus on homeownership (despite some softening of this aspect) and developing integrated human settlements or open towns. To a large extent, post-apartheid policy is the direct opposite of apartheid policy. When put into practice, for example, by demolishing single-sex hostels or transforming them into single-family quarters, the new policy has had severe unintended consequences. Informal settlements have proliferated in mining areas. A choice of housing has not improved the housing conditions of all mineworkers. And, by encouraging place attachment, asset generation and perceived social stability, the new policy puts the new homeowners in a mining town at risk if the mine closes.

Yet very little notice is taken of what mine closure means for housing policy in mining areas. The simplistic assumption in policy and policy guidelines is that closure will be planned, economic diversification is easy and policy therefore does not need to consider linking housing policy and mine closure. There are two inherent assumptions in the current policy: that it is easy to create post-mining economies and that mine downscaling or closure can be seen as a separate issue from housing and settlement policies.

My arguments in this chapter point to the three dependencies discussed in Chapter 1. I see goal dependency in the policy attempts to deal with mine housing's historical problems. Most of the policy documents refer to historical injustices and inhumane practices and are vocal about how to deal with them. They assert that these injustices must be addressed by homeownership, norms and standards and the creation of open towns. Dependence on this goal will create new path dependencies. Launched on the path to stability, a mining town is at risk of becoming stuck when a mine closes: the rigidity that sets in will be disastrous for mineworkers who have bought property. A continuing path dependency, to which policy pays little attention, is the persistence of migrant labour. This path is at odds with the goal of getting mineworkers to settle near their workplace. Finally, the policy focus on housing choice has also changed the interdependencies between government, mining companies, unions and workers. The new housing landscape in many mining towns has made life very different for thousands of mineworkers, and not necessarily better. The following three chapters look at the empirical data that shows how our three case study towns have been affected by these dependencies.

References

Bezuidenhout, A. & Buhlungu, S., 2011. From compounded to fragmented labour: mineworkers and the demise of compounds in South Africa. *Antipode*, 43(2), pp. 237–263.

Burger, P. & Geldenhuys, J., 2018. Work, wages and welfare in Postmasburg. In: L. Marais, P. Burger, & D. Van Rooyen, eds. *Mining and Community in South Africa: From Small Town to Iron Town*. London: Routledge, pp. 173–198.

Charlton, S., 2004. *An Overview of the Housing Policy and Debates, Particularly in Relation to Women*. Johannesburg: Centre for the Study of Violence and Reconciliation.

Charlton, S., 2009. Housing for the nation, the city and the household: competing rationalities as a constraint to reform? *Development Southern Africa*, 26(2), pp. 302–315.
Charlton, S., 2013. *State Ambitions and Peoples' Practices: An Exploration of RDP Housing in Johannesburg*, PhD thesis, University of Sheffield.
Charlton, S. & Kihato, C., 2006. Reaching the poor? An analysis of the influences on the evolution of South Africa's housing programme. In: U. Pillay, R. Tomlinson & J. du Toit, eds. *Democracy and Delivery. Urban Policy in South Africa*. Cape Town: HSRC Press, pp. 252–282.
Cloete, J., 2009. *Housing Options for Mineworkers in Arid and Semi-arid Regions: The Case of Kathu*. Bloemfontein: Unpublished master's degree, University of the Free State.
Cloete, J. & Marais, L., 2009. Mining, aridness and housing: the case of the village under the trees. *Town and Regional Planning*, 55, pp. 31–38.
Cloete, J. & Marais, L., 2021. Mine housing in the South African coalfields: the unforeseen consequences of post-apartheid policy. *Housing Studies*, (accepted). https://doi.org/10.1080/02673037.2020.1769038.
Cloete, J., Venter, A. & Marais, L., 2009. Breaking New Ground, social housing and mineworker housing: the missing link. *Town and Regional Planning*, 54, pp. 27–36.
Cole, M. & Broadhurst, 2021. Measuring the Sustainable Development Goals (SDGs) in mining host communities: a South African case study. *The Extractive Industries and Society*. https://doi.org/10.1016/j.exis.2020.11.012.
Crush, J., 1989a. Accommodating black miners: homeownership on the mines. *South African Review*, 5, pp. 335–347.
Crush, J., 1989b. Migrancy and militance: the case of the National Union of Mineworkers of South Africa. *African Affairs*, 88, pp. 5–13.
Crush, J., 1992. The compound in post-apartheid South Africa. *Geographical Review*, 82(4), pp. 388–400.
Crush, J., 1994. Scripting the compound: power and space in the South African mining industry. *Environment and Planning D: Society and Space*, 12(3), pp. 301–324.
Crush, J. & James, W., 1991. Depopulating the compounds: migrant labor and mine housing in South Africa. *World Development*, 19(4), pp. 301–316.
Daw, O., 2020. Challenges and opportunity of housing for black miner in South Africa. *European Journal of Economics and Business Studies*, 6(1), pp. 28–36.
Demissie, F., 1998. In the shadow of the gold mines: migrancy and mine housing in South Africa. *Housing Studies*, 13(4), pp. 445–469.
Denoon-Stevens, S. & Cloete, J., 2018. Mineworker housing. In: L. Marais, P. Burger & D. Van Rooyen, eds. *Mining and Community in South Africa: From Small Town to Iron Town*. London: Routledge, pp. 141–156.
Department of Housing, 1994. *White Paper: A New Housing Policy and Strategy for South Africa*. Pretoria: Government Printer.
Department of Housing, 2004. *Breaking New Ground: A Comprehensive Plan for the Development of Sustainable Human Settlements*. Pretoria: Department of Housing.
DME (Department of Minerals and Energy), 1998. *White Paper: A Minerals and Mining Policy for South Africa*. Pretoria: Government Printer.
DME (Department of Minerals and Energy), 2002. *Broad-based Socio-economic Empowerment Charter for the South African Mining Industry*. Johannesburg: Co-developed by the Department of Mineral Resources, Chamber of Mines, The South African Development Mining Association.
DME (Department of Minerals and Energy), 2004. *Guidelines for the Development of Social and Labour Plans*. Pretoria: DME.

DMR (Department of Mineral Resources and Energy), 2010a. *Guidelines for the Development of Social and Labour Plans*. Pretoria: DMR.
DMR (Department of Mineral Resources and Energy), 2010b. *Amendment of the Broad-based Socio-economic Empowerment Charter for the South African Mining and Minerals Industry*. Pretoria: DMR.
DMR (Department of Mineral Resources and Energy), 2018. *Broad-based Economic Empowerment Charter for the Mining and Minerals Sector, 2018*. Pretoria: DMR.
DMR (Department of Mineral Resources and Energy), 2019. *Draft Reviewed Housing and Living Conditions Standard for the Minerals Industry, 2019*. Pretoria: DMR.
Drewes, E. & Van Aswegen, M., 2008. Towards a sustainable mining habitat in South Africa. *WIT Transactions on the Ecology and Environment*, 117, pp. 23–32.
Huchzermeyer, M., 2001a. Consent and contradiction: scholarly responses to the capital subsidy model for informal settlement intervention in South Africa. *Urban Forum*, 21(1), pp. 71–106.
Huchzermeyer, M., 2001b. Housing for the poor? Negotiated housing policy in South Africa. *Habitat International*, 25, pp. 303–331.
Huchzermeyer, M., 2001c. Low-income housing and commodified urban segregation in South Africa. In: C. Haferburg & J. Obenbrugge, eds. *Ambiguous Restructuring of Post-apartheid Cape Town*. Humburg: Lit Verlag, pp. 115–131.
Huchzermeyer, M., 2009. The struggle for in situ upgrading of informal settlements: a reflection on cases in Gauteng. *Development Southern Africa*, 26(1), pp. 59–73.
Huchzermeyer, M., 2011. *Cities with 'Slums'. From Informal Settlement Eradication to a Right to the City in Africa*. Cape Town: UCT Press.
Huchzermeyer, M. & Karam, A., 2006. *Informal Settlements: A Perpetual Challenge*. Cape Town: UCT Press.
Laburn-Peart, C., 1990. Homeownership schemes for black mineworkers: overstating and underplanning. *Urban Forum*, 1, pp. 69–82.
Louw, H. & Marais, L., 2018. Mining and municipal finance in Kathu, an open mining town in South Africa. *The Extractive Industries and Society*, 5(3), 278–283.
Mabin, A., 1991. The dynamics of urbanisation since 1960. In: M. Swilling, R. Humphries & K. Shubane, eds. *Apartheid City in Transition*. Cape Town: Oxford University Press, pp. 33–47.
Macmillan, H., 2012. Mining, housing and welfare in South Africa and Zambia: a historical perspective. *Journal of Contemporary African Studies*, 30(4), pp. 539–550.
Manenzhe, T., 2018. *A Critical Review of the Housing Policy and the State's Intervention in Mining Towns in South Africa*. Cape Town: Unpublished master's degree, University of Cape Town.
Marais, L., 2013. The impact of mine downscaling in the Free State Goldfields. *Urban Forum*, 24, pp. 503–521.
Marais, L., 2018. Housing policy in mining towns: issues of race and risk in South Africa. *International Journal of Housing Policy*, 18(2), pp. 335–345.
Marais, L. & Cloete, J., 2013. Labour migration, settlement and mine closure in South Africa. Geography. *Geography*, 98(2), pp. 77–84.
Marais, L. & Cloete, J., 2015. Financed homeownership and the economic downturn in South Africa. *Habitat International*, 50, pp. 261–269.
Marais, L., Denoon-Stevens, S. & Cloete, J., 2020. Mining and urban sprawl in South Africa. *Land Use Policy*, 98, p. 103953.

Marais, L., Haslam McKenzie, F., Nel, E., Deacon, L., van Rooyen, D., & Cloete, J., 2018. The changing nature of mining towns: reflections from Australia, Canada and South Africa. *Land Use Policy*, 76, pp. 779–788.

Marais, L. & Nel, E., 2016. The dangers of growing on gold: lessons from the history of the Free State Goldfields, South Africa. *Local Economy*, 31(1–2), pp. 282–298.

Marais, L., Nel, V., Rani, K., van Rooyen, D., Sesele, K., Van der Watt, P. & Du Plessis, L., 2021a. Economic transitions in South Africa's secondary cities: governing mine closures. *Politics and Governance*, 9(3). (accepted). https://doi.org/10.17645/pag.v9i3.4032.

Marais, L. & Ntema, J., 2013. The upgrading of an informal settlement in South Africa: twenty years onwards. *Habitat International*, 39, pp. 85–95.

Marais, L., Ntema, J., Campbell, M., Cloete, J. & Lenka, J., 2022. Informal settlements in the mining context. In: L. Marais, P. Burger, M. Campbell, S. Denoon-Stevens & D. van Rooyen, eds. *Coal and Energy in Emalahleni, South Africa. Considering a Just Transition*. Edinburgh: Edinburgh University Press, pp. 73–86.

Marais, L., Ntema, J., Cloete, J. & Lenka, M., 2021b. Mine housing, assets and informality in Rustenburg: implications of mine closure. In: L. Marais, M. Campbell, S. Denoon-Stevens & D. van Rooyen, eds. *Mining and Community in the South African Platinum Belt: A Decade after Marikana*. New York: Nova.

Marais, L., Van Rooyen, D., Nel, E. & Lenka, M., 2017. Responses to mine downscaling: evidence from secondary cities in the South African Goldfields. *The Extractive Industries and Society*, 4(1), pp. 163–171.

Marais, L. & Venter, A., 2006. Hating the compound, but … Mineworker housing needs in post-apartheid South Africa. *Africa Insight*, 36(1), pp. 53–62.

Murray, C., 1980. Migrant labour and changing family structure in the rural periphery of Southern Africa. *Journal of Southern African Studies*, 6(2), pp. 136–159.

National Council of Provinces, 2015. *Revitalisation of Distressed Mining Towns Programme: Department of Human Settlements Briefing*. Cape Town: National Council of Provinces briefing.

Nkosi, M., 2017. *Black Workers, White Supervisors: The Emergence of the Labor Structure in South Africa*. Trenton: Africa World Press.

Ntema, J., Marais, L., Cloete, J. & Lenka, M., 2017. Social disruption, mine closure and housing policy: evidence from the Free State Goldfields, South Africa. *Natural Resources Forum*, 41(1), pp. 31–40.

Owen, J., Kemp, D. & Marais, L., 2021. The cost of mining benefits: localising the resource curse hypothesis. *Resources Policy*, 74, 102239.

Pelders, J. & Nelson, G., 2019. Living conditions of mine workers from eight South African mines. *Development Southern Africa*, 36(3), pp. 365–282.

Rubin, M. & Harrison, P., 2016. An uneasy symbiosis: mining and informal settlement in South Africa with particular reference to the Platinum Belt in North West Province. In: L. Cirolia, T. Gorgens, M. van Donk, S. Drimie & W. Smit, eds. *Upgrading Informal Settlements in South Africa. A Partnership-based Approach*. Cape Town: UCT Press, pp. 145–174.

Stewart, T. & Drewes, E., 2018. The Khumani approach to homeownership in Postmasburg. In: L. Marais, P. Burger & D. Van Rooyen, eds. *Mining and Community in South Africa: From Small Town to Iron Town*. London: Routledge, pp. 157–170.

Stuckler, D., Steele, S., Lurie, M. & Basu, S., 2013. 'Dying for gold': the effects of mineral mining on HIV, tuberculosis, silicosis and occupational diseases in southern Africa. *International Journal of Health Services*, 43(4), pp. 639–649.

Tomlinson, M., 1997. *Mortgage Bondage? Financial Institutions and Low-cost Housing Delivery*. Johannesburg: Centre for Policy Studies Research Report.
Tomlinson, M., 1998. South Africa's new housing policy: an assessment of the first two years, 1994–1996. *International Journal for Urban and Regional Research*, 22(1), pp. 137–146.
Tomlinson, M., 1999. From rejection to resignation: beneficiaries' views on the South African Government's new housing subsidy system. *Urban Studies*, 36(8), pp. 1349–1359.
Tomlinson, M., 2002. Efforts and errors: South Africa's search to extend housing finance to low-income households. *Housing Finance Quarterly*, 17(2), pp. 8–14.
Tomlinson, M., 2006. From 'quantity' to 'quality': restructuring South Africa's housing policy ten years after. *International Development Planning Review*, 28(1), pp. 85–103.
Tomlinson, M., 2007. The development of a low-income housing finance sector in South Africa: have we finally found a way forward? *Habitat International*, 31(1), pp. 77–86.
Tshangana, M., 2015. *Selected Committee Presentation on the Revitalisation of Distressed Mining Towns Programme*. Pretoria: Pretoria: National Department of Human Settlements.
Van der Watt, P. & Marais, L., 2019. Normalising mining company towns in Emalahleni, South Africa. *The Extractive Industries and Society*, 6(4), pp. 1205–1214.
Van der Watt, P. & Marais, L., 2021. Implementing social and labour plans in South Africa: reflections on collaborative planning in the mining industry. *Resources Policy*, 71, p. 101984.
Venter, A., Marais, L., Hoekstra, J. & Cloete, J., 2015. Reinterpreting South African housing policy through State Welfare Theory. *Housing, Theory and Society*, 32(2), pp. 346–366.
Wilson, F., 1972. *Migrant Labour: Report to the South African Council of Churches*. Johannesburg: South African Council of Churches.

5 Post-apartheid migrant labour patterns

5.1 Introduction

This is the first of three empirical chapters using data from the household surveys conducted by the University of the Free State in Tsantsabane in 2015, in Emalahleni in 2017 and in Rustenburg in 2018 (see Chapter 1). I analyse post-apartheid migration patterns in the case studies and discuss the implications for mine closure. In Chapter 4, I referred to the historical role of migrant labour in the mining industry. Although there has been a decline in migrant worker numbers, many mineworkers are still migrants. The persistence of migrant labour is an example of a long-term dependency. Yet the current policy ignores migrant labour. Policymakers either try to eliminate it or assume that it does not exist. Worse than that, many policymakers see migrant labour, and mobility generally, as a threat to stability, wealth creation and modernity. Policies therefore promote integrated communities, permanent settlements and homeownership (see Chapter 4).

I argue that, on the contrary, migrant labour has advantages and does not necessarily conflict with wealth creation. It provides flexible access to jobs, allows for a flexible response to mine closure and, as the evidence in this chapter shows, can enable workers to create wealth and accumulate assets. There may be a case for not advocating place attachment but rather promoting the livelihood diversity that mobility and migrant labour offer. Policymakers need to take into account the temporary nature of mining. For many mineworkers, migrant labour is a way to deal with mine closure and avoid the long-term dependency created by homeownership and attachment to a particular place.

I review the international and African literature on mining, mine closure and migrant labour in Section 5.2, and then turn to evidence from South Africa and address the following questions:

- How have mineworker migration trends changed and do these trends differ for the various locations and subsectors of the industry?
- What other migration trends has mining triggered?
- Are migrants wealthier than non-migrants and do they own more assets?
- Which factors contribute to the development of place attachment in mining towns?
- What are the levels of social disruption in mining towns?

DOI: 10.4324/9781003262015-5

5.2 The literature on migration, mining and mine closure

The literature challenges policies that assume permanence. Permanence is valued because it can lead to place attachment, local development, wealth, assets and modernity. Yet, mining is a temporary economic activity. Mining busts are a reminder of this temporariness in the mining industry. The desire for permanence and the reality of impermanence co-exist in mining settlements.

5.2.1 The shift away from company towns

As mining became commercialised over the past 200 years, migration to mining towns increased. As I outlined in Chapter 1, company towns were at first created to attract labour. These towns have had mixed success in managing migration and settlements for mineworkers. On the positive side, mine employees rented mine houses at below market value and received services at reduced rates. Generally, company towns provided a good living environment. When the mine closed, the mining company took responsibility for the town and assisted the employees to relocate peacefully, although there have been instances of post-mining conflict in such towns (Chaloping-March, 2017). The mineworkers could relocate without having property to sell in a low-demand context. On the negative side, the mining companies were heavy-handed in managing company towns, and this has come under scrutiny (Obeng-Odoom, 2018). Too often, the company dominated local decision-making and played the role of a public institution. Regional integration and planning were also difficult.

Over the past 50 years, mining companies have moved away from the company town model in two ways: not developing new company towns, and normalising existing company towns and promoting the development of open towns. This meant that democratic rule and public management became the norm. Elected governance structures became responsible for managing mining booms and busts (Keough, 2015). But often, these structures have struggled to cope with the large influx of people or find appropriate strategies for managing the decline (Van Assche et al., 2020)

Normalisation and open towns emphasised permanence. This placed pressure on governance structures to diversify local economies to ensure long-term economic prosperity (Haslam McKenzie et al., 2009). Diversifying local mining economies was a daunting task because of the usually remote locations of these towns. Dealing with the effects of mining booms and busts was no longer the responsibility of the mining company but that of households and local governance. Normalisation also shifted the viewpoint of policies for mining settlements from seeing these settlements as temporary to emphasising their permanence. The change towards normalisation did bring benefits, however. For example, open towns increased the land tenure options for people working in the mining industry and homeownership became available (Bradbury & St-Martin, 1983).

5.2.2 Fly-in-fly-out arrangements

The change from company towns to open towns was accompanied by the 'fly-in-fly-out' (in some cases, bus-in-bus-out or drive-in-drive-out) arrangement. In this kind of shift work, mineworkers fly in to a mining site, stay in a worker camp for a period of work and then return home for a break. A typical arrangement is two months' work followed by two weeks' break. There is no permanent residence at the mining site for the labour force.

Fly-in-fly-out policies have reduced the number of settlements near the mines (Simard et al., 2019). As Storey (2010, p. 1161) puts it, the 'new town' model has been replaced by the 'no town' model. Robson (1988) noted that in Canada 'fly-in-fly-out commuter camps and short-term residency communities' had become 'the norm for new resource sites of the 1980s'. Jones and Buckley (2017, p. 9) say the Australian outback mining towns 'have now been largely replaced by fly-in-fly-out (FIFO) operations'. Fly-in-fly-out policies have been prominent in Canada since the 1960s (Markey et al., 2016; Golebiowska & Carter, 2017) and in Australia since the 1980s (Saxinger, 2021) and were also being applied in Russia in the 1990s (Saxinger, 2021). Long-distance commuting was introduced in Australia in the 1960s for offshore and onshore oil and gas extraction and has been used by some uranium, gold and diamond mines but not by the coal mines (Hogan & Berry, 2000). Institutionalised migrant labour in South Africa was a long-standing example of a bus-in-bus-out policy. However, unlike migrant labour in South Africa during apartheid, today's fly-in-fly-out policies are for practical and not ideological reasons.

Among the reasons for the development of fly-in-fly-out arrangements are labour force preference (not wanting to settle with their families in remote locations and preferring to live in the coastal cities), cheaper transport costs that make commuting cheaper, business centralisation (single locations), a need to reduce business costs (by not spending on peripheral activities like housing) and changes in the labour regimes, with shift work becoming more popular (Francis & Hille, 2002; Jones & Buckley, 2017; Kozlov et al., 2017; Paredes et al., 2018). Fly-in-fly-out arrangements arose together with increased wages, mechanisation, outsourcing and block roster shifts. Company towns became redundant and existing towns did not necessarily benefit from fly-in-fly-out arrangements. Like the company town model, fly-in-fly-out reinforced temporariness.

Markey et al. (2016) point to the advantages of fly-in-fly-out arrangements. These include reducing the ecological footprint of settlements near mining operations, minimising economic and social disruption at closure and providing lifestyle choice for mineworkers. For example, when a mine closes, the mining company or the local authorities do not have to deal with the sizeable ecological footprint of settlements. Mobility and temporariness have social and ecological benefits and make closure easier by not creating place attachment and localised wealth and assets.

Despite the prominence of fly-in-fly-out arrangements, the system has received criticism (Storey, 2001; Haslam McKenzie et al., 2009; Carrington & Pereira, 2011; Ivanova & Rolfe, 2011; McKenzie et al., 2015; Perry & Rowe, 2015; Haslam McKenzie, 2016; Markey et al., 2016; Packey & Maybee, 2016; Rebar et al., 2018; Simard et al., 2019; Korneeva & Simonova, 2020; Mayes, 2020; Neil & Neis, 2020; Roberts et al., 2021; Saxinger, 2021). Fly-in-fly-out arrangements do not foster local economic development and they minimise the potential local economic benefits of mining. Smaller local workforces with no family members mean less local spending. Research in Canada shows that the lack of links between the mine and the local community has also resulted in governments neglecting these remote places. Overall, fly-in-fly-out arrangements have reduced contact between mineworkers and local people. There are concerns that the mineworkers use some of the local infrastructure (such as roads and health care facilities) but do not pay local taxes. Fly-in-fly-out arrangements have social consequences for workers and their families. These include mental health concerns, worker stress from being disconnected from families, inability to adapt to living arrangements in worker camps and adverse effects on family life. Furthermore, Deacon et al. (2017, p. 95) note that 'the transiency of FIFO workers affects place-attachment and the long-term sustainability of the region'.

Both company towns and fly-in-fly-out policies have attracted criticism. However, the fly-in-fly-out system acknowledges the temporary nature of mining, which should receive more attention in mining policy, especially as regards mine closure. Governments often see mining as the economic sector to build wealth and ensure modernity. However, their mining-linked wealth creation policies seldom consider the temporariness of mining.

5.2.3 *Migration, mining and mine closure in Africa*

Research on mine closure in Africa is rare in a field dominated by research from the Global North (Mfune et al., 2020). The environmental aspects of closure tend to dominate the research, but a few studies have linked migration, mining and mine closure in Africa. I highlight four aspects of this relationship, drawing on the following sources: Ferguson (1999), Lange (2006), Nyame et al. (2009), Larmer (2010), Gough and Yankson (2012), Bryceson and MacKinnon (2012), Obeng-Odoom (2014), Hilson and Garforth (2013), Kotsadam and Tolonen (2016), Jønsson and Bryceson (2017), Gough et al. (2019), Jønsson et al. (2019) and Knierzinger and Sopelle (2019).

First, despite many African countries emphasising permanence, transience and migrant labour dominate the African mining landscape. Many African governments view the mining sector as supporting wealth creation and modernity ideals. The Zambian government, for example, saw the copper mines as the way to modernise Zambia's economy and advance its people. As Ferguson (1999, p. 44) puts it: 'The conventional wisdom written into nearly all existing overview narrations of Copperbelt labour history is that the Copperbelt working class has "developed" through a progressive, linear movement defined by the two poles of labour migration and permanent urbanisation'. But continued labour migration hampered the ideals of modernity and permanent urbanisation. The Zambian government wanted to create social stability by reducing migrant labour and emphasising permanent urban residence. There are several problems with this approach. It assumes long-term mining prosperity, which, in turn, assumes stable global prices and adequate local resources. Booms and busts on the Copperbelt have shown this assumption to be mistaken. The government saw only one way towards modernity: urbanisation. As Ferguson's statement shows, the government polarised permanence and transience: it viewed permanent urban settlement and urbanisation as directly opposed to mobility and migrant labour. It assumed that migrant labour and continued mobility could not create wealth, assets or modernity. In its focus on permanence, it ignored the reality that migrant labour and mobility is a livelihoods approach to deal with the mining industry's booms and busts. Ferguson's (1999) book shows how mineworkers who kept their rural ties and migrated had alternative livelihoods to fall back on when the mine closed.

Second, migration associated with mining in Africa is multifaceted. Bryceson and MacKinnon (2012) show that mining places in sub-Saharan Africa have two primary characteristics: cosmopolitan and temporary. The relationship between mining and migration is less structured in the Global South than in the Global North. In the Global North, mining leads to in-migration and closure to outmigration. Processes in the Global South are more informal. There are examples of high influx into mining areas and mining towns developing and then shrinking when mining ends. However, a few other factors play a role. For Bryceson and MacKinnon (2012, p. 513), referring to Tanzania, the relationship between urbanisation and migration in Africa is about the need to either 'dig deeper' or 'move on'. In Tanzania, this dilemma is usually associated with artisanal mining, but it also seems to be the case in commercial mining. Where there are high levels of mobility, households do not necessarily invest in housing in the mining areas. Many of these Tanzanian mineworkers have housing assets in larger cities or towns. Mobility helps them to accumulate assets where they prefer to invest and avoid the risk of settling near the mine and then seeing it decline. Hilson and Garforth (2013) found that in Ghana having more than one home is a popular concept and mining is used as a livelihood diversification strategy, particularly by households involved in agriculture. They note how artisanal mining is contributing to Ghana's economic transition.

Third, in contrast to the emphasis on the impermanence of mining and the migrant labour option, there is evidence that some mineworkers prefer a considerable degree of permanence. Gough et al. (2019, p. 2670), in a study of workers in a secondary city in Ghana, say that 'contrary to popular understandings of incomers to mining settlements as nomadic opportunists, migrants often aspire to build their own houses and establish a family, which promotes their attachment to these settlements and their desire to remain'. Obeng-Odoom's (2014) study confirms this preference.

And finally, there is evidence that the drive for permanence in mining settlements can have disastrous effects when a mine closes. One of the primary examples of mine decline comes from the copper industry in Zambia. Mfune et al. (2020) found that mine closure in Kabwe, Zambia, resulted from slumping resources prices and the inability of the state-owned mining company to modernise. The consequences for the town were devastating: job losses, inability to find jobs elsewhere, demand for goods and services falling and unplanned settlements mushrooming. Yet, households that had maintained their rural links seemed able to build on existing alternative livelihoods.

The literature from Africa points to the complex nature of the relationship between mining and development. It describes how people deal with the temporariness of mining and how formal towns that grew up because of mining find it difficult to keep up the momentum after mine closure. Mobility becomes an asset to help mineworkers deal with mining decline. However, most policies do not consider the value of migration.

5.3 Migration, mining and mine closure in South Africa

Here we look at how mine closure in South Africa affects migration patterns. I compared migration trends between 1996 and 2016 in municipalities whose mining economy has grown with those where it has declined. I used the 39 municipalities classified as 'secondary cities' in South Africa's intermediate city support programme. (This classification excludes the eight metropolitan areas and numerous smaller urban areas.) Next, I sampled 12 of these 39 that had substantial mining economies and divided them into mining growth municipalities and mining decline municipalities. Mining growth municipalities are those where mining's share of the local economy was three times more than mining's share of the national economy and the mining economy was stable or growing between 1996 and 2016. Seven municipalities fit this definition: Ba-Phalaborwa (copper), Emalahleni (coal), Fetakgomo Tubatse (platinum), Lephalale (coal), Madibeng (platinum), Metsimaholo (coal), Rustenburg (platinum) and Steve Tshwete (coal). Mining decline municipalities are those that experienced a decline in mining's share in the local economy of at least 15% between 1996 and 2016. These are Govan Mbeki (coal), Matjhabeng (gold), Matlosana (gold), Merafong City (gold) and Rand West (gold). Figure 5.1 compares the population and economic indicators for these two types of settlement.

The mining growth municipalities recorded economic growth of 2.2% per annum between 1996 and 2016. The figure for mining decline municipalities was −1.16% per annum, meaning that their economies were on average 28% smaller in 2016 than in 1996. The average annual employment growth in the

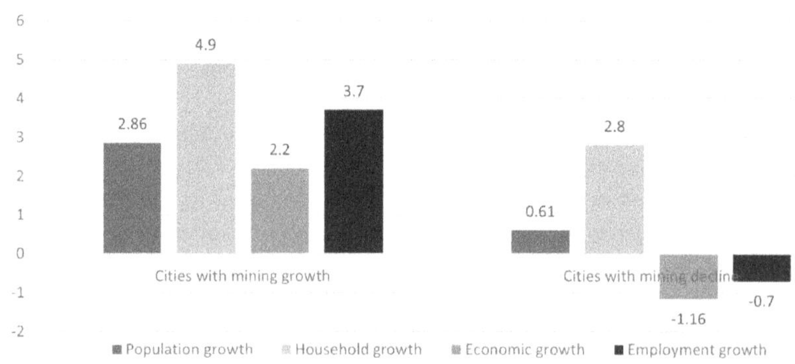

Figure 5.1 Economic and population indicators for growing and declining mining municipalities, 1996–2016.

Sources: Stats SA (1996, 2001, 2013, 2016); Global Insight (2020).

mining growth municipalities was 3.7% for this period, but −0.7% in the mining decline municipalities, where there has been a rapid reduction in the numbers employed on the mines. Only one of the five mining decline municipalities, Govan Mbeki, recorded employment growth as a result of diversifying its economy from mining to manufacturing as a result of the multinational corporation Sasol investing in its Secunda plant. The mining growth municipalities all had employment growth.

Population growth figures in the mining decline municipalities were mixed: two had negative growth and three positive. Their average annual growth was 0.61%. In contrast, all the mining growth municipalities had positive population growth, recording an average annual growth of 2.86%. Compared with the national population growth rate of 1.4% per annum for the period, the mining decline municipalities have had an outflow of people. Every one of the 12 municipalities has seen growth in the number of households, regardless of whether it experienced mining or population growth or decline. The average annual household growth for the mining growth municipalities was 4.99% and for the mining decline municipalities 2.94%.

The contradiction is apparent: the mining decline municipalities experience low and sometimes negative population growth rates, but rapid household growth. There are two main reasons for this. One is that household formation has generally outstripped population growth in South Africa over the past three decades (during which approximately 3.5 million South African households have benefited from housing subsidies). This is partly because of a housing policy that benefits core households instead of extended families or individuals. To access a housing subsidy, a person needs to form a new household. The other reason for this contradiction is the post-apartheid response to the history of migrant labour. Policies promoting open towns and homeownership have encouraged new household formation and urban sprawl (Marais et al., 2020). Abandoning migrant labour also meant that mineworkers could bring their families closer to the mining operations. Sometimes, the new household formation results from the housing policies described above, but in other cases it originates from land invasions or legal land expansions by the municipalities. The irony in the case of the mining decline municipalities is that new household formations have resulted in residential land use expansion despite the rapid economic decline and population outflow. Research in Merafong city (Marais & De Lange, 2021) and Matjhabeng confirms these trends (Marais et al., 2020).

The South African evidence shows that a decline in economic activity leads to a decline in employment and a slowing down of in-migration. In some cases, there is even a decline in the population. Even with a decline in their mining economy, some municipalities experienced a growth in the number of households. Government policies emphasising homeownership and open towns to deal with distress in mining towns have been a prominent contributing factor. It is evident that the mining decline municipalities do not acknowledge the reality of decline. Their overarching goal is to show growth, expand land settlement and provide housing. An acknowledgement of decline is tantamount to accepting failure.

5.4 Migration trends in the three case study areas

In the previous section, we looked at the relationship between economic trends and migration in mining areas in South Africa. We now look at how existing migration trends are related to wealth creation against the potential consequences of mine closure in the three case study areas, Emalahleni, Rustenburg and Tsantsabane. (Emalahleni and Rustenburg were included in the group of seven mining growth municipalities listed above. Tsantsabane is a mining growth municipality but was not included because it is too small to be classified as a 'secondary city'.) The government formally phased out migrant labour and compound living during the 1990s and early 2000s. As I have explained in Chapter 4, the responsibility for housing then shifted from the mining companies to individuals and the local authorities. The overarching policies of open towns, integrated communities and normalisation were intended to counter migration and mobility and create stability through place attachment.

In this section, I discuss the chronology of migration (Figure 5.2) and the migration trends (Tables 5.1 and 5.2). I distinguish between the migration of mining and non-mining households. A mining household is defined as a household with at least one member employed in mining and a non-mining household as a household that has no member employed in mining.

Figure 5.2 shows that the in-migration in all three places started around 1990. Overall, 89% of mining and non-mining household members have migrated since 1990. The increased migration since the 1990s coincides with the start of the mining boom, which peaked in the mid-2000s. In Tsantsabane, however, most migrants have arrived since 2010 as a result of the opening of the Kolomela mine in 2013 and the expansion of the Assmang mine at Beeshoek. Most of them

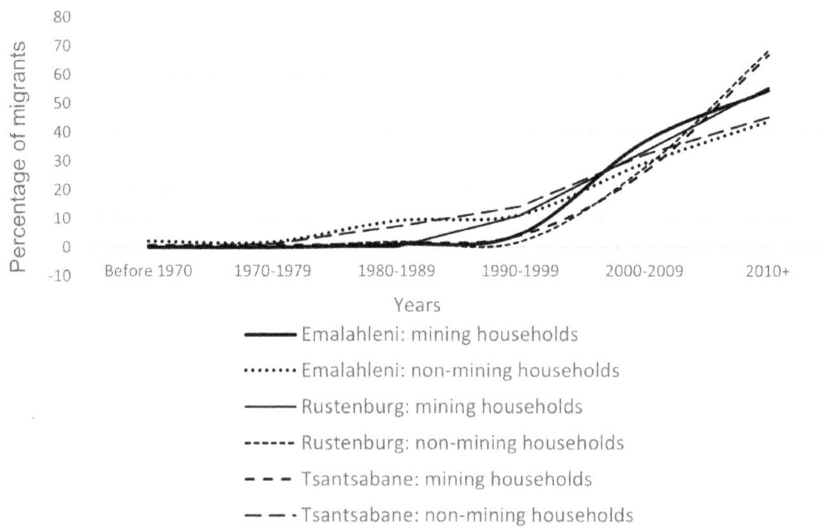

Figure 5.2 Percentage of migrants per annum to the three case study areas.
Source: Household surveys.

migrated to these mining areas for the same reason: to find work or to make a living from the mining economy. In Rustenburg, the in-migration of non-mining household members has been particularly noticeable since 2010, with 68% of them settling in Rustenburg since 2010 compared to 55% of mining household members. Table 5.1 takes the analysis further by profiling the main characteristics of migration in the study areas.

The table points to two contrasting findings: migration continues and yet there is evidence that more mineworkers are living in the place where they work. Approximately 45% of the members of mining households in Rustenburg are from outside the North West and 35% lived elsewhere ten years ago. The corresponding figures for Emalahleni and Tsantsabane are smaller. The percentages of mining-employed household members in the three case studies who lived elsewhere ten years ago are 43% in Tsantsabane, 31% in Emalahleni and 16% in Rustenburg: overall, about 30%. The average percentage of household members living in non-mining households is 21% across the three case studies. Rustenburg's migration is mainly from the Eastern Cape, a historical source of labour for the mines. Some migrants are from the Free State and Gauteng, two provinces which have had declining gold mining economies. For mineworkers who lost their jobs in gold mining in the 1990s, the move to Rustenburg's platinum mines made sense because the mining process is similar in these mines. Migrants to Emalahleni originate mostly from Limpopo and KwaZulu-Natal, the latter also having a historical link with migrant labour. Tsantsabane's migrants are from the Free State and the North West, the latter being a historical migrant labour source. Tsantsabane has the lowest percentage of people from outside its province, the Northern Cape: less than 18% of mining household members are from elsewhere.

Although these figures show high levels of in-migration into all three areas, there is also evidence of local people being employed, which could mean less migration by mineworkers. The figures show that migrant labour continues, but the scale of the migration is far smaller than I expected.

Table 5.2 next assesses the main characteristics of migrant households (a household with at least one migrant), comparing all households with mine-employed migrants and the averages for households in the three areas.

The evidence from Table 5.2 tells us a number of things about migrant households. Households in Emalahleni and Tsantsabane that have a history of migration are considerably smaller than the average. In Rustenburg, they are a little larger (2.8 compared with the average of 2.5). Tsantsabane has the largest average household size, probably because of the rental housing provided by Kumba Iron Ore, which makes it possible for mineworkers to bring their families with them. Mining households earn much more than the average income for all households – migration to the mines is evidently lucrative. The levels of informal housing are fairly high (see Chapter 7 for a more detailed discussion), suggesting that local authorities are struggling to manage the influx of people.

However, the link between migration and informality is not necessarily associated with mineworkers. The percentage of mining households who have migrated and are living in an informal house is much lower than the average for Emalahleni

Table 5.1 Characteristics of migration in the three case study areas, in percentages

Indicator	*Emalahleni Mpumalanga (2017)*	*Rustenburg North West (2018)*	*Tsantsabane Northern Cape (2015)*
Mining HH members born in the province where mining takes place (only mine employment)	74.5	59.7	81.8
Mining HH members born in the province where mining takes place (mine and other employment)	84.1	81.9	91.6
Non-mining HH members born in the province where mining takes place	80.5	84.1	89.5
HHs with nobody employed: members born in the province where mining takes place	78.1	66.4	91.7
Mineworkers born in the province where mining takes place	67.2	54.7	75.1
Mining HH members who lived in another place ten years ago (only mine employment)	25.0	35.4	32.7
Mining HH members who lived in another place ten years ago (mine and other employment)	11.3	30.3	16.8
Non-mining HH members who lived in another place ten years ago	18.1	30.2	15.9
Mineworkers who lived in another place ten years ago	30.7	16.0	42.7
Mining HH members who ever lived in another place (only mine employment)	23.3	20.5	22.5
Mining HH members who ever lived in another place (mine and other employment)	11.8	16.7	12.5
Non-mining HH members who ever lived in another place	19.9	11.6	14.2
HHs with nobody employed: members who ever lived in another place	21.4	14.0	14.1
Mining HH members: two main provinces (other than home province) from which they migrated (only mine employment)	Limpopo KwaZulu-Natal	Eastern Cape Gauteng	North West Free State
Mining HH members: two main provinces (other than home province) from which they migrated (mine and other employment)	Limpopo KwaZulu-Natal	Eastern Cape Free State	North West Free State
Non-mining HH members: two main provinces (other than home province) from which they migrated	Limpopo KwaZulu-Natal	Eastern Cape Gauteng	North West Western Cape
HHs with nobody employed: members born in the province where mining takes place: two main provinces (other than home province) from which they migrated	Limpopo KwaZulu-Natal	Eastern Cape Gauteng	North West Free State

Source: Household surveys.

Note: Calculations were made using the person roster in the household (HH) survey. A mining household is a household with at least one member employed in mining. A non-mining household is a household that has no member employed in mining. A household with nobody employed is a household where none of the members are formally employed.

Table 5.2 Characteristics of migrant households in the three case study areas, Emalahleni, Rustenburg and Tsantsabane

Indicator	*Migrated at any time (all HHs)*			*Migrated in past ten years (all HHs)*			*Migrated at any time (mining HHs)*			*All HHs*		
	E	*R*	*T*	*E*	*R*	*T*	*E*	*R*	*T*	*E*	*R*	*T*
Household size	2	2.5	3.1	1.7	2.4	2.9	1.9	2.8	2.8	2.3	2.5	3.3
Informal house (%)	16.3	21.2	30.4	20.4	23.4	31.5	5.2	12.0	20.2	14.5	26.1	17.9
Happy/satisfied with house (%)	74	63.3	71.6	71.2	59.0	72.9	84.2	72.4	78.5	76	61.0	72.3
Rental housing (%)	63.9	66.2	37.6	76.6	60.6	45.7	73.5	66.7	53.8	54.6	45.9	29.3
Average income (ZAR)	14,160	9,048	7,169	4,406	8,258	7,511	19,056	13,698	11,000	14,357	8,735	6,582
Average wealth self-ranking	3.3	3.5	3.4	3.3	3.7	3.4	3.0	3.2	3.2	3.3	3.8	3.5
Another house elsewhere (%)	5.8	13.0	9.7	6.1	13.4	9.2	1.2	19.2	12.9	4.6	9.7	7.6
Water inside the house (%)	51.2	41.8	59.9	46	45.6	59.8	59.4	57.8	67.9	61.9	47	61.1
Mortgage (% of owners)	12.0	31.9	22.4	26.5	33.3	32.9	15.2	50	33.6	11.0	19.8	19.6
Number of assets	7.2	6.4	7.1	6.8	6.0	6.9	8.5	7.3	8.0	7.8	6.7	7.2
Prefer to own dwelling (% of non-owners)	49.8	50.5	52.7	45.9	45	51.3	52.3	54.2	58.3	50.4	47.8	53.4

Source: Household surveys.
Note: A migrant household is defined as a household with at least one migrant.

and Rustenburg and only a little higher for Rustenburg. Many non-mineworkers who flock to the mining areas searching for work are living in informal houses. A larger percentage of migrants have houses elsewhere. Although it is difficult to determine the relationship with this second home, it shows that housing assets exist elsewhere. Households will have to decide where to invest in housing – near the mine or in the place where they have the second house. Mining households that have migrated have more assets than the average for all households. However, this is not generally the case for all migrants, as many non-mining migrants are poor and in search of any jobs they can find in mining areas. As with the income figures, the high levels of assets for mineworkers who have migrated shows that migration can go hand in hand with asset creation. Policymakers should not assume that migration either prevents or supports asset-building. Some people need to find homeownership and stability, but others like to invest in a house elsewhere. Finally, although the self-ranking of wealth for most migrants was slightly lower than the average, the differences were minuscule and not worth building any specific argument on (this point is discussed further below).

Households that migrated differ slightly from the full sample of households. For example, they are smaller than the average for all households. Migration ensures a livelihood for some of them. Earning an income and building assets do not necessarily depend on stability and homeownership. These can be achieved by migration. But migration could require different policy approaches and policymakers will need to understand that migrant households may build their housing assets in locations other than the mining areas.

Figure 5.3 shows how migrant households rank their wealth, currently and at specific periods in the past. The point is to see whether they think their wealth is increasing or decreasing. The questionnaire asked them to rank their wealth on a scale of one to six for their current situation, five years ago, ten years ago and when the respondent was aged 15. In Tsantsabane, the questionnaire did not include five years ago.

In all three case study areas, migration increases the wealth ranking. Mining households that have migrated have seen rapid increases in their wealth in Emalahleni and Rustenburg. Coal mining in Emalahleni is responsible for the highest increases in wealth ranking for the three areas.

The evidence presented in this section has several implications. Mining booms and in-migration go hand-in-glove. This is not new, and much of the international literature has focused on the social disruption caused by mining booms (see Chapter 2). In this book, I focus on the dependence that booms can create. They foster place attachment and encourage both mining and non-mining households to accumulate assets. They create thriving environments where households and local governments do not consider the likelihood of busts. Policymakers typically assume that it is place attachment and homeownership that create stability and wealth. However, the evidence above shows that migration can create assets and increase wealth ranking. Stability and minimal migration are not necessarily prerequisites for prosperity. The evidence suggests that South Africa's policies promoting homeownership and stability are working. The vital question is, however: who is most at risk when a mine closes, the homeowner or the migrant?

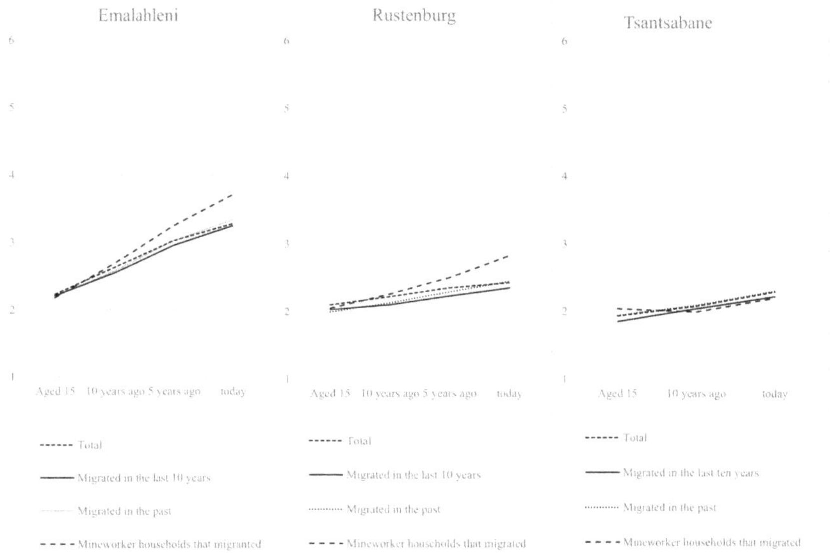

Figure 5.3 Migrant households' self-ranking of wealth on a six-step ladder.

Note: 'Today' means at the time of the survey: 2015 for Tsantsabane, 2017 for Emalahleni and 2018 for Rustenburg.

5.5 Migration and place attachment

5.5.1 Overview

Two contrasting trends are visible in the migration patterns discussed above: continued migrant labour and increased settlement near the mines. There is also evidence of increased wealth and assets among mineworkers, despite migration. This section looks at the respondents' degree of place attachment and perceptions of social disruption. To investigate place attachment, we asked respondents how likely they were to remain in their current town. They could pick one of five answers: a strong preference to stay, a moderate preference to stay, unsure, a moderate preference to leave, a strong preference to leave.

5.5.2 Factors contributing to place attachment

We start with an overview of place attachment in the three case study areas (see Table 5.3).

Table 5.3 shows that respondents had a strong preference for staying where they were, particularly in Tsantsabane, where 79% of respondents strongly preferred to stay in the area, compared with Emalahleni's 64% and Rustenburg's 59%. This preference is generally lower for non-mineworkers, mostly people who have moved into all three areas hoping for a job because of the mining developments. The

Table 5.3 Preference to stay in the area (per household, in percentages)

Strong or moderate preference to stay in the area	*Emalahleni*	*Rustenburg*	*Tsantsabane*
Total population	63.8	58.9	78.6
Non-mining household	47.4	59.8	66.5
Mining households	70.0	56.6	77.4
Mix of mining and non-mining employment	87.7	81.8	77.7
No formal employment	42.4	56.9	84.2
Employed	67.6	60.2	78.8
Migrated in the past ten years	54.6	49.8	73.3
Ever migrated	58.9	55.5	77.2
Never migrated	71.0	63.0	84.5

Source: Household surveys

mineworkers' higher preference for staying indicates the development of place attachment.

The literature on place attachment and its contributing factors was reviewed in Chapter 2. Elsewhere, I have analysed the place attachment of mineworkers in Emalahleni and Rustenburg (Marais et al., 2021). That study found that, contrary to the findings of the international literature, homeownership among mineworkers did not contribute to place attachment, but higher levels of satisfaction with the house did contribute. In some cases, older and less educated mineworkers also had less place attachment (probably the result of historical migrant labour). To expand on Marais et al. (2021) and understand the role of various factors in the development of place attachment in mining settlements, I conducted a linear regression of place attachment for households in all three study areas and a set of selected variables. The full linear regression analysis is attached as Annexure 5.1, and I discuss only the main findings below (see Table 5.4).

Two indicators contributed to place attachment in all three areas: ownership (rather than rental) and being happy or satisfied with the house. Marais et al. (2021) found only the latter significant. The finding regarding ownership corroborates international empirical research (see Chapter 2). In this chapter's analysis of all three areas, I did not find that mine employment created place attachment, which shows how mobility remains important for mining households.

Indicators that were statistically significant in two of the three places were being older than 40 (Emalahleni and Rustenburg) and having a larger house (Rustenburg and Tsantsabane). The age factor probably means those respondents had lived longer in the area, as migrants are generally younger, and the larger house factor supports levels of satisfaction with the house. The age finding for the full sample contrasts with the finding by Marais et al. (2021) that older mineworkers in Rustenburg had lower levels of place attachment.

Statistically significant factors only applicable in Emalahleni were higher ranking on the asset index, having mine and non-mine employment rather than no formal employment and having no formal employment rather than having mine employment. The first two of these supports some of the other indicators that are

Annexure 5.1 Linear regression of place attachment and selected variables

Model summary	*Emalahleni*				*Rustenburg*				*Tsantsabane*			
n	887				889				624			
Predicted accurately	65.1				59.2				80.3			
Model chi square	301.9				328.5				74.6			
Model p	0.000				0.000				0.000			
Cox & Snell R square	0.288				0.309				0.113			
Nagelkerke R square	0.397				0.417				0.179			
Variables	B	OR	SE	p	B	OR	SE	p	B	OR	SE	p
Constant	−1.408	0.245	0.483	0.004	−1.078	0.340	0.395	0.006	1.510	4.527	0.723	0.037
40 years or younger	−0.852	0.427	0.247	0.001	−0.512	0.599	0.198	0.010	−0.456	0.634	0.380	0.229
Asset index (/15)	0.140	1.151	0.050	0.005	0.069	1.071	0.042	0.103	−0.068	0.935	0.052	0.192
Number of rooms	0.033	1.034	0.060	0.580	0.256	1.292	0.059	0.000	0.193	1.213	0.068	0.005
Female	0.131	1.140	0.191	0.493	0.051	1.052	0.174	0.772	−0.125	0.883	0.231	0.590
Married/cohabiting	−0.340	0.712	0.195	0.081	−0.035	0.966	0.183	0.849	0.273	1.314	0.290	0.347
Matric or higher	0.259	1.296	0.217	0.233	−0.038	0.962	0.185	0.836	−0.307	0.736	0.236	0.194
All resident ten years ago	−0.161	0.851	0.206	0.433	0.262	1.299	0.187	0.162	0.550	1.733	0.243	0.024
Only non-mine employment	0.555	1.741	0.299	0.064	−0.189	0.828	0.231	0.413	−0.151	0.860	0.362	0.677
Mixed employment	1.451	4.267	0.421	0.001	0.201	1.222	0.439	0.648	−0.253	0.776	0.497	0.610
Only mine employment	0.957	2.605	0.333	0.004	−0.373	0.689	0.216	0.084	0.008	1.008	0.383	0.983
Happy or satisfied with housing	1.296	3.653	0.215	0.000	1.777	5.910	0.177	0.000	0.999	2.715	0.263	0.000
Backyard	−0.231	0.794	0.228	0.312	0.529	1.698	0.243	0.029	−0.306	0.736	0.467	0.511
Rental	−0.664	0.515	0.249	0.008	−0.522	0.593	0.216	0.016	−0.761	0.467	0.287	0.008
Rent free	−0.236	0.790	0.352	0.503	−0.572	0.564	0.262	0.029	−0.670	0.512	0.388	0.084
Informal housing	−0.075	0.928	0.345	0.829	−0.784	0.457	0.287	0.006	0.863	2.370	0.469	0.066
Water in house	0.433	1.542	0.242	0.074	0.216	1.242	0.224	0.334	−0.511	0.600	0.298	0.086
Flush toilet	0.270	1.310	0.337	0.423	−0.482	0.618	0.244	0.049	−0.457	0.633	0.512	0.372
Average or above self-ranking of wealth	−0.135	0.874	0.228	0.555	0.701	2.016	0.304	0.021	0.215	1.239	0.266	0.420
House repairs in the past two years	−0.291	0.748	0.233	0.211	0.099	1.105	0.223	0.656	−0.652	0.521	0.254	0.010

Source: Household surveys.
Notes: B = beta, OR = odds ratio, SE = standard error. Statistically significant results are shaded.

Table 5.4 Factors significantly related to the development of place attachment

Indicator	*Emalahleni*	*Rustenburg*	*Tsantsabane*
40 years or younger	Older than 40 years (rather than younger)	Older than 40 years (rather than younger)	None
Asset index (/15)	Having more assets	None	None
Number of rooms	None	Larger house (more rooms)	Larger house (more rooms)
Female	None	None	None
Married/cohabiting	None	None	None
Matric or higher	None	None	None
All residents here ten years ago	None	None	All residents here ten years ago
Non-mining households	None	None	None
A mix of mining and non-mining households	Mining and non-mining employment (rather than households with no employment)	None	None
Mining households	No formal employment (rather than only mine employment)	None	None
Happy or satisfied with housing	Happy or satisfied with housing (rather than not)	Happy or satisfied with housing (rather than not)	Happy or satisfied with housing (rather than not)
Backyard	None	Backyard unit (rather than own stand)	None
Rental	Ownership (rather than rental)	Ownership (rather than rental or rent free)	Ownership (rather than rental)
Rent-free	None	None	None
Informal housing	None	Formal housing (rather than informal)	None
Water in house	None	None	None
Flush toilet	None	No flush toilet	None
Average or above-average self-ranking of wealth	None	Average or above-average self-ranking of wealth	None
House repairs in the past two years	None	None	No house repairs in the past two years

Source: Household surveys.

significant. It was not very clear how to interpret the third one. It could indicate that mine employment does not automatically lead to place attachment because of its migratory nature.

In Rustenburg, a backyard unit as opposed to a house in a separate stand and a formal house as opposed to an informal house were statistically significant. An average or above-average socio-economic self-ranking was positively related to place attachment. The importance of a backyard unit instead of a house on a formal stand is probably the result of a high degree of informality for houses on separate stands in Rustenburg. Strangely, not having a flush toilet was also related to place attachment. Again, I am unable to provide an adequate explanation.

In Tsantsabane, place attachment was significantly associated with all members of the household having lived in the areas ten years ago and with not having done repairs in the past two years. The length of time spent in one place is a common factor contributing to place attachment. But the finding about repairs is counter-intuitive and not easy to explain.

The statistically significant results generally match the international empirical evidence. Ownership, strong satisfaction with the house, length of stay, house size, resident's age and a formal house are significantly related to place attachment in at least one of the three case study areas. The place attachment created by mining means dependency for mining households, with the attendant risks if the mine closes: job losses, inability to pay rent or mortgages, poverty. However, we should not underestimate the interconnectedness of mining and non-mining households; mining also creates dependencies for non-mining households.

5.5.3 *Place attachment and wealth*

The linear regression in Table 5.4 shows that in Emalahleni possessing more assets and in Rustenburg a higher self-ranking of wealth were significantly related to place attachment. Figure 5.4 provides more details in this respect.

In Emalahleni and Rustenburg, respondents with a strong preference to stay had a higher self-ranking of wealth and more assets. In Tsantsabane, those with a strong or moderate preference to stay and those with a strong or moderate preference to leave both had the same self-ranking of wealth, and those who preferred to leave had more assets than those who preferred to stay. I refer to this difference for Tsantsabane again in Chapters 6 and 7. This finding is the direct result of Kumba's mine rental housing programme in Tsantsabane that has provided good quality housing. The findings again challenge policymakers' assumption that assets and wealth creation cannot co-exist with migration.

5.6 Evidence of social disruption

I do not have data for possible social disruption because of mine closure, as mine closure has not occurred on any substantial scale in any of the three case study areas. However, investigating social disruption associated with mining *growth* in these areas could give us an idea of the implications of mine closure. Here, we look at the levels of disruption and the types of households affected by it.

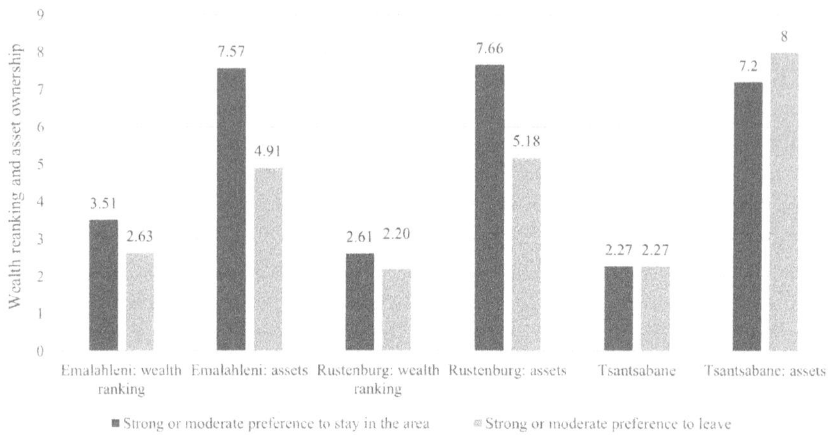

Figure 5.4 High and moderate preference to stay in the area (per household) linked to wealth ranking and asset ownership.

Source: Household surveys.

5.6.1 *Distrust as a measure of social disruption*

To test people's level of trust in neighbours and strangers and thus their perception of the stability of their area, in the survey's follow-up interviews respondents were asked what the likelihood would be of a wallet that they had lost containing ZAR 200 being returned by someone close by or a stranger. Table 5.5 shows the percentages of respondents who thought it unlikely that the wallet would be returned. We are interested particularly in the differences between the households that had migrated and those that had not.

Emalahleni respondents who had migrated in the past ten years or ever migrated were more likely than the average for their area, and more likely than those who had never migrated, to consider the return of the wallet improbable, whether by someone close by or a stranger. In that town, 73% of all respondents and 65% of the non-migrants said that someone close by would not return the wallet, but 81% of those who had migrated in the past ten years and 79% of those who had ever migrated were of this opinion. In Rustenburg, distrust of neighbours was high for all categories, but slightly higher for migrants. Migrants in these two areas seem to experience a higher level of social disruption, if distrust of one's neighbour is any indication. We do not see this pattern in Tsantsabane. In that town, 78% of the whole sample thought the wallet would not be returned by someone close by, but the percentages for migrant households were 67% and 75%. They were markedly more suspicious of strangers. In all three areas, households where all members were employed in mining were more inclined to believe that someone close by would return the wallet.

We might wonder what makes Tsantsabane different. For a start, it is much smaller than the other two places (see Chapter 1). Its good quality housing (both

Table 5.5 Likelihood that a wallet would be returned (in percentages of respondents who thought it unlikely)

Not likely at all that it would be returned by…	*Emalahleni*		*Rustenburg*		*Tsantsabane*	
	SCB	*Stranger*	SCB	*Stranger*	SCB	*Stranger*
Total population	73.3	90.5	86.3	94.9	78.0	92.2
Non-mining households	86.0	87.3	86.4	97.0	81.1	92.6
Mining households	65.9	85.2	83.2	93.3	72.4	90.7
Mix of mining and non-mining households	61.9	90.8	79.2	88.9	85.9	93.6
No formal employment	91.4	97.6	90.4	97.0	77.3	93.5
Employed	30.4	88.5	84.3	94.3	79.5	91.9
Migrated in past ten years	80.9	94.1	86.7	94.0	67.3	92.5
Ever migrated	79.1	94.5	85.4	94.0	74.9	91.4
Never migrated	64.6	84.8	77.4	95.8	82.7	93.2

Source: Household surveys.
Note: SCB = someone close by.

rented and owned), provided by the mines (Kumba and Assmang, respectively), creates a sense of belonging and community. Good housing can minimise social disruption and I think it has played a role in Tsantsabane, helping to minimise the social disruption that migrants can experience (see Chapter 6).

5.6.2 *Factors affecting perceptions of crime*

To assess people's experience of crime in the three study areas, the survey asked respondents to rate their perceptions of the levels of six types of crime in their area: burglaries, muggings or thefts; violence between members of the same household; violence between members of different households; gangsterism; murder, shootings or stabbings; and drug or alcohol abuse. A crime index was developed by rating the responses on a scale of 1 to 5, with 4 being 'fairly common' and 5 'very common'. The average was then calculated by summing the responses for each of the six types of crime.

To analyse in more detail the factors affecting perceptions of crime, I developed a linear regression of the crime index (see Annexure 5.2). Table 5.6 shows the main results of this regression.

Two factors in particular were statistically significantly related to the crime index in all three case study areas: being unhappy or unsatisfied with the house, and having a flush toilet. This contrasts with the high levels of place attachment associated with being happy or satisfied with the house (see Section 5.5.1). Being happy or satisfied with the house created place attachment, but being unhappy or unsatisfied with it was related to experiencing crime. The number of rooms in a house was also significant in all three places. Having fewer rooms was significantly related to perceptions of crime in Rustenburg and Tsantsabane, but in Emalahleni having more rooms was significantly related.

Annexure 5.2 Linear regression of a crime index with selected variables

	Emalahleni				*Rustenburg*				*Tsantsabane*			
n	884				859				627			
Model F	14.0				15.0				8.8			
Model p	0.000				0.000				0.000			
R square	0.236				0.253				0.216			
Adjusted R square	0.219				0.236				0.192			
	B	SE	t	p	B	SE	t	p	B	SE	t	p
Constant	24.950	1.272	19.613	0.000	29.946	1.756	17.054	0.000	24.336	2.476	9.831	0.000
40 years or younger	0.049	0.353	0.138	0.890	0.416	0.507	0.822	0.411	0.762	0.760	1.003	0.316
Asset index (/15)	−0.185	0.080	−2.318	0.021	−0.373	0.113	−3.305	0.001	−0.230	0.121	−1.904	0.057
Number of rooms	0.230	0.088	2.620	0.009	−0.905	0.141	−6.424	0.000	−0.397	0.149	−2.660	0.008
Female	−0.482	0.284	−1.696	0.090	0.297	0.451	0.658	0.510	0.235	0.515	0.457	0.648
Married/cohabiting	0.557	0.294	1.896	0.058	−0.706	0.477	−1.481	0.139	−1.704	0.643	−2.651	0.008
Matric or higher	−1.773	0.330	−5.366	0.000	−1.789	0.486	−3.683	0.000	−0.478	0.541	−0.883	0.378
All resident ten years ago	0.448	0.314	1.429	0.153	0.873	0.497	1.757	0.079	1.137	0.549	2.073	0.039
Only non-mine employment	−0.764	0.463	−1.652	0.099	0.019	0.602	0.031	0.975	−1.207	0.760	−1.589	0.113
Mixed employment	−2.943	0.586	−5.024	0.000	−1.093	1.028	−1.063	0.288	−2.203	1.094	−2.013	0.045
Only mine employment	−2.373	0.513	−4.630	0.000	−1.216	0.560	−2.172	0.030	−1.915	0.808	−2.369	0.018
Happy or satisfied with housing	−1.590	0.369	−4.312	0.000	−1.836	0.484	−3.794	0.000	−1.871	0.591	−3.164	0.002
Backyard	−0.440	0.340	−1.293	0.196	−2.462	0.679	−3.624	0.000	1.656	1.168	1.417	0.157
Rental	0.587	0.362	1.622	0.105	−2.958	0.557	−5.313	0.000	−2.110	0.670	−3.150	0.002
Rent-free	1.792	0.510	3.515	0.000	−0.852	0.659	−1.292	0.197	−1.602	0.923	−1.735	0.083
Informal housing	−1.273	0.561	−2.270	0.023	−0.774	0.786	−0.985	0.325	−0.709	1.035	−0.685	0.493
Water in house	−0.641	0.408	−1.571	0.117	−0.665	0.618	−1.076	0.282	−2.045	0.639	−3.200	0.001
Flush toilet	1.869	0.574	3.257	0.001	1.913	0.667	2.870	0.004	3.511	1.118	3.141	0.002
Average or above self-ranking of wealth	−0.620	0.365	−1.698	0.090	−1.175	0.724	−1.623	0.105	1.355	0.581	2.330	0.020
House repairs in the past two years	1.367	0.342	4.000	0.000	−0.324	0.559	−0.581	0.562	1.998	0.574	3.478	0.001

Source: Household surveys.

Notes: B = beta, OR = odds ratio, SE = standard error. Statistically significant results are shaded.

To assess people's experience of crime in the three study areas, the survey asked respondents to rate their perceptions of the levels of six types of crime in their area: burglaries, muggings or thefts; violence between members of the same household; violence between members of different households; gangsterism; murder, shootings or stabbings; and drug or alcohol abuse. A crime index was developed by rating the responses on a scale of 1 to 5, with 4 being 'fairly common' and 5 'very common'. The average was then calculated by summing the responses for each of the six types of crime.

Table 5.6 Factors significantly related to perceptions of crime in the area

Indicator	Emalahleni	Rustenburg	Tsantsabane
40 years or younger	None	None	None
Asset index (/15)	Fewer assets	Fewer assets	None
Number of rooms	More rooms	Fewer rooms	Fewer rooms
Female	None	None	None
Married/cohabiting	None	None	Not married or cohabiting
Matric or higher	Education below matric (rather than matric or higher)	Education below matric (rather than matric or higher)	None
All residents ten years ago	None	None	All residents ten years ago
Non-mining households	None	None	None
Mixed of mining and non-mining households	No formal employment (rather than mixed or mine employment)	None	No employment (rather than mixed or mine employment)
Mining households	No employment (rather than mine employment)	No employment (rather than mine employment)	None
Happy or satisfied with housing	Unhappy with house (rather than happy or satisfied)	Unhappy with house (rather than happy or satisfied)	Unhappy with house (rather than happy or satisfied)
Backyard	None	House on separate stand (rather than backyard dwelling)	None
Rental	None	None	Owning the house (rather than rental)
Rent-free	Rent-free (rather than owning)	Rent-free (rather than owning)	None
Informal housing	Formal housing (rather than informal)	None	None
Water in house	None	None	No water in house
Flush toilet	Flush toilet (rather than not)	Flush toilet (rather than not)	Flush toilet (rather than not)
Average or above self-ranking of wealth	None	None	Average or above-average self-ranking of wealth
House repairs in the past two years	Home repairs in the past two years	None	Home repairs in the past two years

Source: Household surveys.

In Emalahleni and Rustenburg, not having matric was significantly related to higher perceptions of crime. In Tsantsabane this did not play a role, but households where no one was formally employed did. In Emalahleni and Rustenburg, perceptions of crime were significantly related to no formal employment rather than mine employment and to occupying a stand rent-free instead of owning. In Emalahleni and Tsantsabane, perceptions of crime were significantly related to no formal employment rather than to mixed or mine employment and to having done house repairs in the past two years. I can suggest no clear explanation why house repairs should be related to perceptions of crime, except that perhaps the repairs were intended to prevent crime or that doing repairs points to a level of place attachment. In Emalahleni, living in formal housing was related to higher perceptions of crime and, like house repairs factor, could point to higher levels of place attachment.

Several indicators were statistically significant for only one area. In Tsantsabane, for example, higher perceptions of crime were related to not being married or cohabitating, having lived there ten years ago, owning a house rather than renting, not having water in the house and an average or above-average self-ranking of wealth. The Tsantsabane profile shows the effect of mining growth on the existing non-mining residents. The effect is stronger here, as the mining influx has only occurred in the past ten years.

Overall, two main points are important. The finding that unhappiness with the house heightens perceptions of crime supports the finding that being happy with the house makes people want to stay in the area (see Section 5.5.2). And importantly for our topic in this chapter, there is no evidence that people who migrate have heightened perceptions of crime levels.

5.6.3 *Crime statistics*

In this section, I triangulate the findings above with crime data from the three case study areas. Working with crime statistics is difficult because crime definitions often change (Marais et al., 2022). Essentially, six main types of crimes are recorded at police stations in South Africa. Figure 5.5 shows the data for these over ten years, 2009–2018, per 100,000 of the population.

In Emalahleni and Rustenburg the crime levels have decreased considerably, but in Tsantsabane the levels of most of the crimes have increased. This is understandable as the main mining operations only started in 2013 and the literature shows that a mining boom brings social disruption. The volatility in the crime levels in Tsantsabane is also noteworthy and provides further evidence of the role of crime in social disruption in the area. From the follow-up interviews conducted in Tsantsabane, it is clear that a substantial portion of this crime is drug-related. In Emalahleni and Rustenburg, the reason for the decrease in the crime levels may be that mining has been prominent in those areas for about three decades. The social disruption has abated over time.

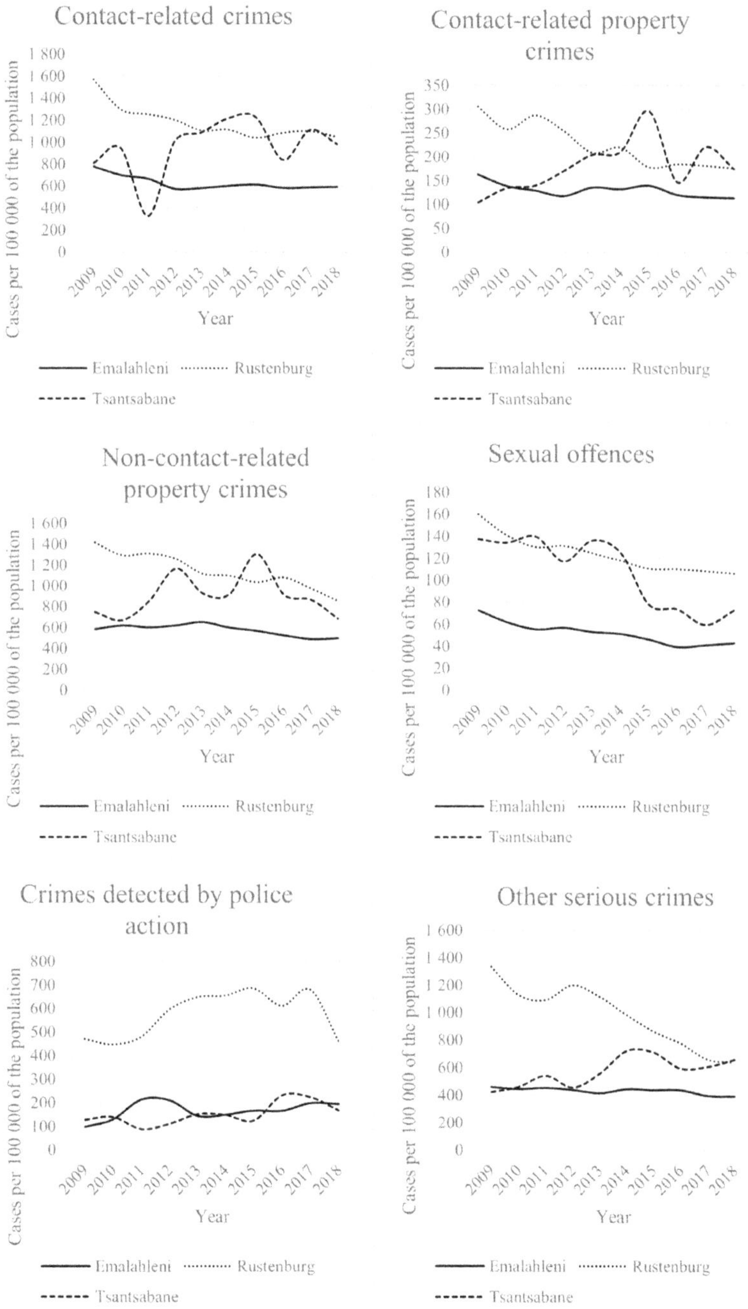

Figure 5.5 Crime trends in the three case study areas, 2009–2018.
Source: Household surveys.

5.7 Conclusion

This chapter assessed migration and mobility in the three case study areas against the background of the South African government's drive to link mining and development to ensure stability (see Chapter 4). Many African governments also use mining to create stability, wealth and modernity. In this development model, migration and mobility are seen as counterproductive. They intend mining to help urbanise the population, build assets and wealth and create place attachment to ensure stability. Continued mobility or migrant labour stands in contrast to this intention – something that policymakers should avoid. The chapter's analysis of current mobility and migrant labour provides evidence that the stability of homeownership is not the only way to create assets and wealth. Migration and mobility are ways to manage livelihood diversification and deal with the likelihood of mine closure. Migrants *can* accumulate assets.

The chapter presented five main findings. First, there is evidence that despite government policies to reduce migrant labour, a fair number of mineworkers continue to migrate. Mineworker migration today differs slightly from historical patterns and is based on individual choice. Ironically, changing labour regimes in the mining industry, like shift work and block roster shifts, have fuelled migration. Government policies that promote homeownership do not necessarily cater for migrant mineworkers. The results have been households living in informal or rental housing in the mining areas. Ignoring mobility and continued migrant labour levels or seeing it as a contradiction to asset and wealth generation is a mistake.

Second, migration and mobility help people to gain wealth and assets. This finding contradicts the assumption in policies in Africa and South Africa. Migrant mineworkers in our surveys recorded more assets and higher wealth rankings than those who did not migrate. Although migrant labour, like fly-in-fly-out arrangements, can have adverse social implications, the evidence shows that wealth is merely created in different places instead of only in the settlement near the mine. In Australia, property markets in coastal cities benefit from fly-in-fly-out. The evidence from Tanzania has also demonstrated that some mineworkers avoid investing in towns near the mines. Their investments go to larger towns and cities.

Third, the linear regression revealed several factors contributing to place attachment in mining towns. The two most significant factors in all three case study areas are homeownership and being happy or satisfied with the house. In a previous study, however, focusing only on mineworkers, homeownership did not feature as a factor contributing to place attachment (Marais et al., 2021). The results in this chapter show no statistically significant relationship between place attachment and mine work. Although there was some evidence of place attachment developing among mineworkers, the lack of a significant relationship indicates that mobility and migrant labour continue in the mining context.

Fourth, although there is evidence that residents who are homeowners and residents who are unsatisfied with their houses tend to have heightened perceptions of crime levels, there is no evidence that this is the case for mineworker migrants.

Finally, the difference between Tsantsabane and the other two case study areas is important. It receives attention later in the book (Chapters 6 and 7). Migrant mineworkers in Tsantsabane had more assets, ranked their wealth the same as other mineworkers or residents and did not perceive crime levels as higher. This is a result of heavy investment in mine rental housing by Kumba Iron Ore in Tsantsabane.

What are the implications for mine closure? Some mineworkers clearly migrate to deal with complex family or livelihood reasons, and there is also evidence that some are simply continuing historical trends. These households would deal with mining decline easily. Kumba's investment in rental housing seems to be a practical way to support flexible livelihood creation and build on the value of mobility and migration. In the case of closure, these mineworkers can return to their houses elsewhere, from which they can migrate to an area where new jobs are available. Yet these realities are seldom considered in policies and programmes. Government policies focusing on homeownership to create stability and place attachment may be putting mining households at risk and making them vulnerable when a mine closes.

References

Bradbury, J. & St-Martin, I., 1983. Winding-down in a Quebec mining town: a case study of Schefferville. *Canadian Geographer*, 27(2), pp. 128–144.

Bryceson, D. & MacKinnon, D., 2012. Eureka and beyond: mining's impact on African urbanisation. *Journal of Contemporary African Studies*, 30(4), pp. 513–537.

Carrington, K. & Pereira, M., 2011. Assessing the social impacts of the resources boom on rural communities. *Rural Society*, 21(1), pp. 2–20.

Chaloping-March, M., 2017. *Social Terrains of Mine Closure in the Philippines*. London: Routledge.

Deacon, L., Papineau, J. & Lamanes, T., 2017. Transiency, fly-in-fly-out workers, and sustainability: perceptions from within a resource-based community. *WIT Transactions on Ecology and the Environment*, 226(1), pp. 95–105.

Ferguson, J., 1999. *Expectations of Modernity: Myths and Meanings of Urban Life on the Zambian Copperbelt*. Berkeley: University of California Press.

Francis, R. & Hille, S., 2002. Setting best practice: FIFO at the Mount Keith concentrator. *Australasian Institute of Mining and Metallurgy Publication Series*, 2, pp. 129–141.

Global Insight, 2020. *Rex Explorer Database*. Pretoria: Global Insight.

Golebiowska, K. & Carter, T., 2017. International immigration trends and data. In: D. Carson, R. Rasmussen, P. Ensign, L. Huskey & A. Taylor, ed. *Demography at the Edge: Remote Human Populations in Developed Nations*. London: Routledge, pp. 53–84.

Gough, K. & Yankson, P., 2012. Exploring the connections: mining and urbanisation in Ghana. *Journal of Contemporary African Studies*, 30(4), pp. 651–668.

Gough, K., Yankson, P. & Esson, J., 2019. Migration, housing and attachment in urban gold mining settlements. *Urban Studies*, 56(13), pp. 2670–2587.

Haslam McKenzie, F., 2016. The socio-economic impacts of long-distance commuting on people and communities. In: F. Haslam McKenzie, ed. *Labour Force Mobility in the Australian Resources Industry: Socio-economic and Regional Impacts*, Heidelberg: Springer, pp. 11–28.

Haslam McKenzie, F., Phillips, R., Rowley, S. & Brereton, D.B.-J.C., 2009. *Housing Market Dynamics in Resource Boom Towns*. Perth: AHURI Final Report.

Hilson, G. & Garforth, C., 2013. 'Everyone now is concentrating on the mining': drivers and implications of rural economic transition in the Eastern Region of Ghana. *The Journal of Development Studies*, 49(3), pp. 348–363.

Hogan, L. & Berry, P., 2000. Mining and regional Australia: some implications of long distance commuting. *Australian Commodities*, 7(4), pp. 648–659.

Ivanova, G. & Rolfe, J., 2011. Using input-output analysis to estimate the impact of a coal industry expansion on regional and local economies. *Impact Assessment and Project Appraisal*, 29(4), pp. 277–288.

Jones, R. & Buckley, A., 2017. From the horse and cart to the internet: a century of rural connectivity change in rural Western Australia. *Bulletin de la Société Géographique de Liège*, 69(2), pp. 9–16.

Jønsson, J. & Bryceson, D., 2017. Beyond the artisanal mining site: migration, housing capital accumulation and indirect urbanisation in East Africa. *Journal of Eastern African Studies*, 11(1), pp. 3–23.

Jønsson, J., Bryceson, D., Kinabo, C. & Shand, M., 2019. Getting grounded? Miners' migration, housing and urban settlement in Tanzania, 1980–2012. *Extractive Industries and Society*, 6(3), pp. 948–959.

Keough, S., 2015. Planning for growth in a natural resource boomtown: challenges for urban planners in Fort McMurray, Alberta. *Urban Geography*, 36(8), p. 2269.

Knierzinger, J. & Sopelle, I., 2019. Mine closure from below: transformative movements in two shrinking west African mining towns. *Extractive Industries and Society*, 6, pp. 145–154.

Korneeva, Y. & Simonova, N., 2020. Job stress and working capacity among fly-in-fly-out workers in the oil and gas extraction industries in the Arctic. *International Journal of Environmental Research and Public Health*, 17, p. 7759.

Kotsadam, A. & Tolonen, A., 2016. African mining, gender, and local employment. *World Development*, 83, pp. 325–339.

Kozlov, A., Gutman, S.R.E. & Zaychenko, I., 2017. Human and economic factors of long-distance commuting technology: analysis of Arctic practices. *Advances in Intelligent Systems and Computing*, 487, pp. 409–420.

Lange, S., 2006. *Benefit Streams from Mining in Tanzania: Case Studies from Geita and Mererani*. Bergen: CMI reports.

Larmer, M., 2010. Historical perspectives on Zambia's mining booms and busts. In: A. Fraser & M. Larmer, eds. *Zambia, Mining and Neoliberalism: Boom and Bust on the Globalised Copperbelt*. New York: Palgrave MacMillan, pp. 31–58.

Marais, L. & De Lange, A., 2021. Anticipating and planning for mine closure in South Africa. *Futures*, 125, p. 102669.

Marais, L., Denoon-Stevens, S. & Cloete, J., 2020. Mining and urban sprawl in South Africa. *Land Use Policy*, 98, p. 103953.

Marais, L., Ndaguba, M., Mmbadi, E., Lenka, M. & Cloete, J., 2022. Mine closure, social disruption and crime in South Africa. *The Geographical Journal*, DOI:10.1111/geoj.12430.

Marais, L., Owen, J., Kotze, T., Nel, P., Lenka, M. & Cloete, J., 2021. Determinants of place attachment among mineworkers: evidence from South Africa. *The Extractive Industries and Society*, 8(3), p. 100943.

Markey, S., Storey, K. & Heisler, K., 2016. Fly-in/fly-out resource developments: implications for community and regional development. In: L. Huskey, P. Ensign, R. Ole Rasmussen, eds. *Demography at the Edge: Remote Human Populations in Developed Nations*. London: Routledge, pp. 213–236.

Mayes, R., 2020. Mobility, temporality, and social reproduction: everyday rhythms of the 'FIFO family' in the Australian Mining Sector. *Gender, Place and Culture*, 27, pp. 126–146.

McKenzie, F., Haslam McKenzie, F. & Hoath, A., 2015. Fly-in/fly-out, flexibility and the future: does becoming a regional FIFO source community present opportunity or burden? *Geographical Research*, 52(4), pp. 430–441.

Mfune, O., Chansa-Kabali, T., Chisola, C. & Manchisi, J., 2020. The legacy of mine closure in Kabwe, Zambia: what can resilience thinking offer to the mining sustainability discourse? In: G. Fasparatos, M. Naidoo, A. Ahmed, A. Karanja & K. Fukushi eds. *Sustainability Challenges in Sub-Saharan Africa: Insights from Eastern and South Africa.* Singapore: Springer Nature, pp. 167–188.

Neil, K. & Neis, B., 2020. 'Mobility has always been a part of my life': work-related mobility and families in Canada. *Canadian Studies in Population*, 47, pp. 111–118.

Nyame, F., Andrew Grant, J. & Yakovleva, N., 2009. Perspectives on migration patterns in Ghana's mining industry. *Resources Policy*, 34(2), pp. 6–11.

Obeng-Odoom, F., 2014. *Oiling the Urban Economy: Land, Labour, Capital and the State in Sekondi-Takoradi, Ghana.* London: Routledge.

Obeng-Odoom, F., 2018. Transnational corporations and urban development. *The American Journal of Economics and Sociology*, 77(2), pp. 447–510.

Packey, D. & Maybee, B. 2016. Economic impact analysis in a resource industry boom. In: F. Haslam McKenzie, ed. *Labour Force Mobility in the Australian Resources Industries: Socio-economic and Regional Impacts.* Singapore: Springer, pp. 81–102.

Paredes, D., Soto, J. & Fleming, D., 2018. Wage compensation for fly-in/fly-out and drive-in/drive-out commuters. *Papers in Regional Science*, 97(4), pp. 1337–1353.

Perry, M. & Rowe, J., 2015. Fly-in, fly-out, drive-in, drive-out: the Australian mining boom and its impacts on the local economy. *Local Economy*, 30(1), pp. 139–148.

Rebar, A., Alfrey, K., Gardner, B. & Vandelanotte, C., 2018. Health behaviours of Australian fly-in, fly-out workers and partners during on-shift and off-shift days: an ecological momentary assessment study. *BMJ Open*, 8(12), p. e023631.

Roberts, C., Darroch, F., Giles, A. & van Bruggen, R., 2021. Plan A, Plan B, and Plan C-OVID-19: adaptations for fly-in and fly-out mental health providers during COVID-19. *International Journal of Circumpolar Health*, 80(1), p. 1935133.

Robson, R., 1988. Wilderness suburbs: boom and gloom on the prairies, 1945–1986. *Prairie Forum*, 13(2), pp. 191–219.

Saxinger, G., 2021. Rootedness along the way: meaningful sociality in petroleum and mining mobile worker camps. *Mobilities*, 16(2), pp. 192–211.

Simard, M., Maltais, E. & Brisson, C., 2019. Fly-in/fly-out in northern Quebec. An exploratory study respecting workers' representations. *espace populations sociétés*, 1, p. 8536.

Stats SA, 1996. *Census Data 1996.* Pretoria: Stats SA.

Stats SA, 2001. *Census Data 2001.* Pretoria: Stats SA.

Stats SA, 2013. *Census Data 2011.* Pretoria: Stats SA.

Stats SA, 2016. *Community Survey 2016.* Pretoria: Stats SA.

Storey, K., 2001. Fly-in/fly-out and fly-over: mining and regional development in Western Australia. *Australian Geographer*, 32(2), pp. 133–148.

Storey, K., 2010. Fly-in/fly-out: implications for community sustainability. *Sustainability*, 2(5), pp. 1161–1181.

Van Assche, K., Verschraegen, G. & Gruezmacher, M. (2020). Taming the boom and the bust? Land use tools for mitigating ups and downs in communities. *Land Use Policy*, 93, 104058.

6 Mining, housing and wealth creation

6.1 Introduction

The previous chapter discussed the relationship between mobility, assets and wealth creation. I noted that post-apartheid policy fosters the link between mining, homeownership, wealth creation and modernity. I showed that despite this emphasis in policy, mineworker mobility can create household assets and wealth. Generally, homeownership and being happy with a house create place attachment, but for mineworkers place attachment is not statistically significantly related to owning a house.

This chapter analyses the nature and scale of housing in the three case study areas: Emalahleni, Rustenburg and Tsantsabane. The analysis determines the risks for the different housing types when a mine closes. South African policy wants mining to help people accumulate housing assets through homeownership. The focus on housing is evident in many African countries. Yet, mine closure holds substantial risks for people who own a house in a mining town. The African literature on mine closure points to the dependencies mining creates and the risks at closure.

I make two main arguments in this chapter. First, I argue that the link between mining, housing, wealth creation and modernity creates dependencies. There is evidence that mining wealth creates local assets and household benefits. But these benefits can come at a cost (Owen et al., 2021). Ownership and excessive focus on local development make households and mining towns overdependent on mining. Mining households and local governments assume that mining brings long-term economic growth. Mine closure does not feature in this mindset: it is easier not to acknowledge any inconveniences that may ensue. But Knierzinger and Sopelle (2019, p. 6) note that mining seldom brings long-term economic prosperity: 'the "reality" of African mining is also one of recurring deceptions'. Second, I argue that appropriate rental housing can help minimise dependence. The empirical evidence shows that rental housing does not prevent asset creation, and it can give households flexibility and mobility at mine closure. Furthermore, the example of mine-supplied rental housing in Tsantsabane provides proof of the value of rental housing provided by mining companies.

DOI: 10.4324/9781003262015-6

6.2 An African perspective on mining, housing and wealth

Mining is an important economic sector for many African states. Although its contribution to gross domestic product (GDP) is modest, it produces foreign exchange. Many of them want to capitalise on the link between mining, housing and wealth creation. The volatility in the mining industry and the threat of mine closure hold much risk for mining settlements, but the African research pays little attention to this (Jønsson & Bryceson, 2017; Jønsson et al., 2019).

The Africa Mining Vision, a policy framework set up by the African Union in 2009, wants the continent to benefit from mining. The South African government has the same aim. Yet this envisaged benefit does not take into account the likelihood of closure. Although there has been an increase in case studies about mining towns in decline, Bradbury and Martin's (1983, p. 128) comment that 'Little attention has been paid to processes of decline or to the phase of uncertainty evident in settlements where decline is taking place' is still applicable today. Mine decline and closure has deferred the modernist dream in Africa (Ferguson, 1999; Knierzinger & Sopelle, 2019). Knierzinger and Sopelle (2019) identify four main characteristics of African mining towns: peripheral location and operating outside the global knowledge flows, environmental degradation, adverse health impacts and high levels of unemployment. These characteristics hinder the dream of modernity and make the effects of closure more severe. They note that mine closure is 'more difficult and conflictual than in high-income countries' and that 'many negative consequences could only be avoided by differently planning and constructing mining towns from the beginning' (2019, p. 146). There is a need to reconceptualise sustainable mining to take account of decline and closure (Mfune et al., 2020).

6.2.1 The Africa Mining Vision minerals governance framework

The Africa Mining Vision wants African countries to integrate mining into countries' economies so they can benefit from mining. It provides an 'overarching framework for achieving inclusive, sustainable mineral-based structural transformation' (African Union, 2009, p. v). The assumption is that mining has not benefited African economies and that more appropriate strategies would foster such development. The Africa Mining Vision is described as 'an innovative and holistic vision for governing the mineral sector on the continent', whose goal is to 'ensure transparent, equitable and optimal exploitation of mineral resources to underpin broad-based sustainable growth and socioeconomic development' (African Union, 2009, p. 1).

The documents set out policy frameworks at the country level and do not pay attention to the subnational level of mining towns. They do, however, refer to social impacts and mining communities. For example, they say there should be

'Improved and sustainable quality of life for mining-affected communities and the country as a whole' (African Union, 2009, p. 6). The Africa Mining Vision assumes that mining should provide local benefits and improve the quality of life of mining-affected communities. While accepting the finite nature of mineral resources, it proposes that better management is required to deal with this reality. It puts much emphasis on managing closure, but mainly in reference to environmental aspects and land rehabilitation.

Since the release of the Africa Mining Vision, a growing body of research has reflected on it (Busia & Akong, 2017; Cawood, 2017; Sturman et al., 2020; Hilson, 2020; Dauda, 2020; Kamlonger, 2020; Béland et al., 2021). The assessments vary: some see it as a foundation for mining; others assess it more critically. The critics question the potential role of the Africa Mining Vision in artisanal mining. They argue in favour of more local participation, emphasising implementation and understanding power relationships. Yet, none of the researchers reflects on the way mining can create local dependencies and the debates seldom consider mine closure. The Africa Mining Vision emphasises the relationship between mining and modernity. Although there are some arguments in favour of this relationship, I believe it is a one-directional approach. Mine decline and closure can break this link. A more appropriate mining vision for Africa should take cognisance of the risks that mining brings to local communities. It should emphasise identifying risks early and minimising the long-term dependence on an industry with a limited life span.

6.2.2 *Mine closure in Africa*

In Chapter 3, I challenged the front-end approach by mining companies that results in the neglect of closure. In research in Ghana, Knierzinger and Sopelle (2019) found that at closure the 'willingness of the companies to make concessions was at its lowest'. An increasing body of African research investigates how mine decline and closure affect the surrounding communities (Larmer, 2006; Nyame et al., 2009; Frazer, 2010; Gewald & Soeters, 2010; Rubbers, 2010; Gough & Yankson, 2012; Kamete, 2012; Bryceson & MacKinnon, 2013; Lee, 2014; Littlewood, 2014; Jønsson & Bryceson, 2017; Knierzinger, 2017, 2018; Gough et al., 2019; Knierzinger & Sopelle, 2019; Mfune et al., 2020; Bowman et al., 2021). Here, I summarise important findings from that literature, which includes case studies from the Democratic Republic of the Congo, Ghana, Namibia, Tanzania, Zambia and Zimbabwe.

Bryceson and MacKinnon (2013) emphasise the temporary nature of mining and the consequences for mineworkers and their families. This means that governments should be careful not to develop long-term plans and strategies on the back of the mining industry.

The link between mining and housing creates long-term dependencies. Policies like the Africa Mining Vision assume that mining wealth will last forever. Many African countries anticipated that mining would bring modernity. Mining would promote urbanisation, create assets, develop urban place attachment and ensure modernity through urban housing. Formal housing would minimise migrant labour and create stability. Zambia initially promoted both migrant labour and stabilisation, but after the 1960s, the focus shifted to urban stabilisation based on the mining economy. Proper towns would have to be created for mineworkers and their families, to minimise migrant labour. In Kabwe, Zambia, the mining company delivered more than 2,000 houses and owned about 30% of the properties in town (Mfune et al., 2020). Mineworkers benefited from relatively high wages and invested in houses and cars and took advantage of the availability of shops to buy luxury goods. When mining collapsed, they lost their incomes and risked losing their assets and houses.

One consequence of decline is that mines transfer the ownership of houses to mineworkers who cannot pay mortgage and municipal services. In addition, decline may create 'parallel informal settlements' in a mining town (Littlewood, 2014, p. 37). Ferguson (1999, p. 13) describes this dependence on the prosperity brought by booms in the Zambian Copper Belt:

> Access to the first-class things of the world—cars, suits, fine clothes, a decent necktie—was not something to look forward to in an anticipated future, but something from a prosperous past—a past now gone, gone never to return and history, as it were, running in reverse.

Dependence can be created by welfare systems set up by the mines. In addition to becoming used to prosperity, the urban system becomes used to welfare programmes and unable to take responsibility for welfare and health. Many mining companies used to operate a cradle-to-grave welfare system, providing subsidised housing, free education, electricity, water and transport and even medicine for mineworkers and their families. Once decay set in, the companies outsourced these services or transferred them to the relevant government departments, many of which were ill-equipped to provide the services. Buses that transported mineworkers and schoolchildren no longer operated, free housing became paid-for housing, free medical treatment vanished and the mines no longer provided food for their workers. In some of these towns, the mining companies were managing the infrastructure. Closure meant that they discontinued this non-core function and water and electricity interruptions became common. Municipalities had to take over and maintain infrastructure, but were under strain because households' ability to pay for water and electricity had diminished because of job losses. To add to the problems, a declining and ageing population needed care, because the young and able workforce is the first to leave these towns. The research reports a deterioration of the quality of health services and even food shortages.

Local businesses that had become dependent on mining had to adapt when decline set in. Businesses that had been selling on credit found they could not collect these debts during the decline. Knierzinger and Sopelle (2019) report a rapid decline in business turnover after closure, with one-third of business operators saying their revenue was only 20% of their pre-mine closure figures. The research noted lower private care use, less eating out and lower shopping volumes over weekends.

The education system also had to adjust, with the private schools experiencing a substantial drop in pupil numbers (although the numbers quickly picked up again). The local people did not have sufficient savings to continue paying their children's school fees and teachers left the mining areas. The rebranding of private schools (with new mission statements to attract pupils) and new 'product lines' in education were common responses to mine closure.

There are also social consequences of mine decline and closure. A decline has been seen in paternal authority, with male mineworkers losing their power over the household and the extended family. However, some authors warn that not too much must be read into these changes, as men remained dominant in their families. Many mineworkers abandoned their families, resulting in 'mass depression' because the 'dream of modernity did not materialise' (Knierzinger, 2018, p. 87). Mining growth had made it possible for them to support not only their immediate family but also the second family at the mines. Closure had serious effects on all these dependents. In Ghana, the decline prompted the formation of an association of second wives, who agitated for their rights as closure also meant a decline in their financial situations.

Much of the research emphasises the way mining creates a dependence that traps households when prosperity wanes. Some researchers, however, have noted the mineworkers' agency in trying to avoid the negative effects of closure. In Tanzania, mineworkers tend to invest in their housing in larger towns and not in the towns near the mines. Jønsson and Bryceson (2017, p. 3) say that 'In anticipation of declining mineral yields and retirement from days of roughing it in mining sites' mineworkers 'endeavour to channel savings into housing in more urbanised locations, aiming to diversify into profitable business activities, living a life with better physical and social amenities'. This point is important. It shows that the workers can understand the temporary nature of mining and use their mobility to benefit from mining without becoming trapped by it.

Fears that a mining town may become a ghost town or a cemetery are commonly mentioned in this body of research. Many local strategies emphasise the importance of economic diversification and the development of local entrepreneurship to prevent these disasters. These attempts have been criticised for continuing neoliberal and paternalistic approaches initiated during colonialism. This seems harsh, but it is a valid criticism of simplistic local strategies for economic diversification. The regional economies in which mining operates are often small and remote, making it difficult to create an alternative economic sector. In Chapter 3, I emphasised the importance of seeing the value of decline. Planning for a shrinking city is an alternative to futile attempts to diversify the economy.

6.3 Housing in the case study areas

Here, I analyse the housing approaches in the three case study areas, describe their housing profiles and discuss a cluster analysis that integrates the three datasets.

6.3.1 Housing approaches

Each of the three case study areas had different ways of housing mineworkers. Up to the 1990s, Emalahleni had company towns called 'mining villages'. These were located next to the collieries and, in the apartheid years, housed only the white mineworkers. Single-sex compounds provided housing for black mineworkers. Three changes have occurred since the early 1990s. The mining companies closed their mining villages, thus reducing their long-term non-mining liabilities. With local government, they set up a comprehensive homeownership system to lure mineworkers to the formal urban areas, from which they would commute to the collieries (bus-in-bus-out) instead of living next to them (Van der Watt & Marais, 2019; Cloete & Marais, 2021). In one case, the government normalised a company town, Rietspruit; a move which, though initially lauded as best practice, showed a deterioration of services in the long term and a low appetite for homeownership (Filitz, 2011). Informal housing increased. There was a rush for rental housing, much of it in the form of small backyard dwellings, and the municipality found itself ill-prepared to deal with the influx and the extra pressure on infrastructure.

In Rustenburg the mines constructed new houses for their mineworkers (Marais et al., 2021). This involved a partnership agreement with private developers to build the houses on behalf of the mining companies. The companies took ownership once the contractors had completed the houses but generally sold them to the workers. A few companies in Rustenburg still own houses, but individual homeownership is common. The companies therefore have very few long-term liabilities and homeownership has become the dominant tenure model for middle- and high-income mineworkers. Despite these efforts by the mines, informal housing has expanded, mainly on tribal land near the mines. Living in informal housing near the mines enables mineworkers to reduce their transport costs. However, the survey found that most of the informal housing in Rustenburg is occupied by low-income migrants not necessarily working in the mines. The informal housing on tribal land has hampered the ability of the municipality to service the land (Ntema, 2019).

Mine housing in Tsantsabane started with the company town at Beeshoek. Originally, there was also a single-sex compound, but the mine phased it out in the early 1990s. The opening of the Kolomela mine about 25 km outside of the town of Postmasburg in 2013 and the closure of the Beeshoek company town were major turning points for Tsantsabane. The town found itself having to provide housing for the mineworkers. Two mining companies, Kumba Iron Ore and Assmang, and the Tsantsabane Local Municipality set up a public-private partnership to deliver houses (Stewart & Drewes, 2018). Assmang provided a capital subsidy to mineworkers who wished to buy houses from the companies, reducing the initial amount that had to be financed. Most homeowners bought their houses with a leasehold agreement, which means that transfer only occurs at the final instalment, until which time the house remains the property of the mining company. Kumba, the owners of Kolomela, provided rental houses to their employees, an exception to current practice. In the remainder of this chapter, I highlight the importance and value of this approach by Kumba. Both Kumba and Assmang employees could also buy houses on the open market.

Emalahleni, Rustenburg and Tsantsabane illustrate the different approaches that mining towns have taken to housing. In all three cases, there are examples of mine housing being transferred to individuals or company towns being closed in line with government policies – shifting the responsibility for housing mineworkers from the mining companies to households and the local municipality. But there are also examples of mining companies taking responsibility for the long-term implications, such as Kumba's focus on rental housing and Assmang's taking co-responsibility for their ownership model. The different approaches have implications for how local governments should respond. For example, an increase in backyard rentals affects infrastructure and land use regulations. Local governments need to deal with the growing informal settlements.

6.3.2 *The changing housing profiles*

Table 6.1 shows the housing profiles from the three surveys in the case study areas. I use this data and the historical data from Statistics South Africa to discuss the current housing.

Two aspects stand out from Table 6.1. First, 25% of the households in the Emalahleni sample were living in a house in a backyard. Although this is partly the result of the sampling, it does indicate a large rental housing market associated with mining. The percentages for backyard housing in Tsantsabane and Rustenburg are much smaller at 9% and 2%. Second, large percentages in Rustenburg and Tsantsabane were in informal housing, which is sometimes a form of rental housing. A large percentage of informal and backyard houses is characteristic of mining towns (Chapter 7 elaborates on informal housing). To put these findings in perspective, Table 6.2 compares the level of informal housing in the three case studies with the rest of South Africa.

In Rustenburg, 29% of households live in informal housing (either in a backyard or on a separate stand). The corresponding figures in Tsantsabane and Emalahleni are 18% and 23%. The average of 24% for the three areas in 2016 is substantially higher than the South African average of 13% in 2016. These high percentages of informal housing show how difficult it is for a

Table 6.1 Housing profiles for three case studies

Type of house	*Emalahleni*		*Rustenburg*		*Tsantsabane*	
	n	%	*n*	%	*n*	%
House or brick structure on a separate stand	438	46.8	696	69.3	623	65.9
Traditional dwelling/hut/structure made of traditional materials	2	0.2	1	0.1	1	0.1
Apartment in a block of flats	30	3.2	4	0.4	38	4.0
Town/cluster/semi-detached house (simplex, duplex, triplex)	22	2.4	16	1.6	9	1.0
House/flat/room in backyard	231	24.7	21	2.1	80	8.5
Informal dwelling/shack in backyard	62	6.6	24	2.4	61	6.5
Informal dwelling/shack not in backyard, e.g. in an informal/ squatter settlement or on farm	71	7.6	225	22.4	108	11.4
Room/flatlet	78	8.3	6	0.6	24	2.5

Source: Household surveys.

municipality to plan and manage the influx of people. Traditional land tenure, which is less strictly regulated, has facilitated the development of informal housing in Rustenburg. Table 6.2 also provides figures for housing in five municipal areas whose mining economies have been shrinking (see Chapter 5). In these municipalities with mine decline, the percentage of informal housing has decreased, but the percentage of formal housing has grown largely in line with the three case study areas. Marais and De Lange (2021) have shown that this increase in formal housing results from government policy. They question the merit of such an approach in places with negative economic growth.

The 6% or 7% of informal housing shown in Table 6.1 for Emalahleni is probably an undercount, if we consider the official statistics in Table 6.2, which show it as consistently in the vicinity of 23% since 1996. The high percentages of people living in informal housing and in backyards of formal houses indicate how difficult it is for a mining municipality to plan for growth when the influx of people outweighs its ability to prepare. The percentages also suggest that people with a choice do not necessarily want to settle permanently in a mining city, but prefer an informal dwelling or cheap backyard renting in a formal area. Yet, municipalities often fail to acknowledge backyard rentals, considering them inappropriate, putting pressure on land and services.

There has also been a substantial growth of formal housing. In Emalahleni and Rustenburg, it has grown rapidly because of mining, and in the latter it has grown faster than informal housing since 2001. However, the growth is also a result of government policies that link mining, housing and wealth creation. Mining

Table 6.2 Formal and informal housing in Emalahleni, Rustenburg and Tsantsabane compared with South Africa as a whole, 1996–2016

	Formal houses	*Informal houses*	*Total houses*	*% informal house on the erf*	*% annual growth of formal houses*	*% annual growth of informal houses*
Emalahleni, 1996	39,634	12,901	55,687	23.2	n.a.	
Emalahleni, 2001	50,254	19,514	74,917	26.0	4.9	8.6
Emalahleni, 2011	92,597	23,138	119,874	19.3	6.3	1.7
Emalahleni, 2016	111,914	34,845	150,420	23.2	3.9	8.5
Rustenburg, 1996	44,398	28,637	74,874	38.2	n.a.	
Rustenburg, 2001	64,820	45,740	113,394	40.3	7.9	9.8
Rustenburg, 2011	136,732	59,268	199,044	29.8	7.7	5.3
Rustenburg, 2016	178,941	76,062	262,576	29.0	5.5	2.5
Tsantsabane, 1996	4,873	784	6,003	13.1	n.a.	
Tsantsabane, 2001	5,535	1,002	6,800	14.7	2.6	5.0
Tsantsabane, 2011	7,061	2,536	9,839	25.8	2.5	9.7
Tsantsabane, 2016	9,203	2,327	11,821	19.7	5.4	−1.7
South Africa, 1996	5,834,819	1,644,388	8,967,512	18.3	n.a.	
South Africa, 2001	7,680,421	1,654,787	11,205,705	14.8	5.7	0.1
South Africa, 2011	11,219,247	1,139,916	14,450,161	7.9	3.9	−3.7
South Africa, 2016	13,404,199	2,193,968	16,923,309	13.0	3.6	14.0
Municipalities with mine decline[a], 1996	200,998	126,156	327,154	38.5	n.a.	
Municipalities with mine decline[a], 2016	480,644	96,726	577,370	16.7	4.5	−1.3

Sources: Stats SA (1996, 2001, 2016).

a The five municipal areas that form part of this group of cities are the Govan Mbeki (coal), Matjhabeng (gold), Matlosana (gold), Merafong (gold) and Rand West (gold). See Chapter 5. The average contribution of mining to their economies dropped from over 65% in 1996 to 31% in 2016. They experienced negative economic growth between 1996 and 2016.

companies have provided extensive homeownership and formal housing incentives in both places

6.3.3 Cluster analysis

To analyse the housing typology further, I conducted a cluster analysis by placing all the households from the three case studies in one sample. There are four main housing types in mining towns. I discuss in detail the characteristics of each cluster (see Table 6.3).

Cluster 1, about 27% of the sample, consists of non-mineworkers living mainly in formal houses and on separate stands. Emalahleni has the highest percentage of respondents in this group, at 34%. This cluster are long-time residents but generally view themselves as poorer than the mining households in Cluster 2. Ownership in Cluster 1 is only 51%, but this is the highest in the four clusters, and considerably higher than the 36% for rental housing. This cluster has the second-highest percentage of respondents unhappy with the house, but generally they have adequate services. About one in four respondents expressed a moderate to strong preference to leave the area, the second-lowest percentage for the four clusters.

Cluster 2, the largest group, about 34% of the sample, is made up mostly of mineworkers who have opted for family houses. Like Cluster 1, this cluster

Table 6.3 Cluster analysis for households in the three case study areas

Indicators	*Cluster 1* (%)	*Cluster 2* (%)	*Cluster 3* (%)	*Cluster 4* (%)
Mining households	2	82	95	21
Formal house	93	99	90	28
House in backyard	13	1	66	22
Renting	36	44	66	27
Owning	51	48	25	29
Always lived here	54	44	37	44
Unhappy with the house	35	14	26	59
Did repairs in the past two years	17	39	20	17
Have a house elsewhere	6	10	7	4
Water in the house	66	96	28	3
Flush toilet	99	99	88	4
Electricity	96	98	95	30
Moderate/strong preference to leave area	24	12	25	35
Above or much above average income	7	21	13	5
Distribution in Emalahleni	20	36	26	18
Distribution in Rustenburg	34	30	13	24
Distribution in Tsantsabane	28	37	13	23
Average distribution	27.3	34.3	17.3	21.7

Source: Household surveys.

contains both owners and renters. The clustering here is mainly because of the mining link, but the cluster remains diverse regarding land tenure arrangements, with owners and renters nearly evenly distributed. Almost half are owners. This cluster has the highest percentage of income above or much above the average. This is directly related to their mining incomes. They have the lowest levels of unhappiness with the house and the highest levels of repairs done. This cluster has the highest percentage with houses elsewhere, which shows the migratory nature of the respondents in this cluster. However, they also have the lowest rate of preference to leave the area. This contradictory finding suggests that some mining households develop place attachment and that the government drive to create adequate housing is reaping benefits.

Cluster 3, about 17% of the sample, is also a cluster made up mostly of mineworkers, but these ones prefer rental housing in backyard units. It is a distinct cluster that points to the demand for rental housing, but households in this cluster tend to rent formal backyard dwellings instead of having family houses like those in Cluster 2. This cluster has the lowest percentage of people who have always lived in these mining towns, which points to current migration levels. Furthermore, 7% have houses elsewhere. The service levels are considerably lower than in Clusters 1 and 2 and one in four respondents expressed an average or high preference to leave the area, the second-highest percentage for any of the clusters. They have the second-highest incomes, with 13% being above or much above average.

Cluster 4, about 22% of the sample, is a cluster of low-income households (mainly non-mineworkers) living in informal houses with the lowest levels of services. These are largely migrant individuals and households. Consequently, they also have the largest percentage with a moderate or strong preference to leave the area.

The difference in place attachment levels between Cluster 2 and Cluster 3 is important as mineworkers dominate both groups. The moderate or strong preference to leave the area in Cluster 3 is 25% compared to the 12% for Cluster 2. This finding shows that renters in less family-oriented mine housing have a lower level of place attachment, contributing to their decision to rent rather than own housing in the three case study areas. The 35% of respondents in Cluster 4 who indicated a moderate or high preference to leave the area point to this group being highly mobile and unlikely to invest in their local housing. This group also has the lowest percentage of households who have formal housing (28%).

6.3.4 *Implications for mine closure*

There are two main conclusions from the cluster analysis. First, mining in South Africa has created a dual formal and informal housing system, as in many mining towns on the continent. Both types are the result of mining growth. The formal system develops because of mine work and government policies are promoting it. Although the formal system sometimes outgrows the informal housing system, the pace of formal housing delivery is seldom fast enough to deal with the influx of people. Second, a rental housing market has developed in all three of the case study areas. This requires more analysis. In Emalahleni, this market is in formal backyards or second buildings on an erf. Rustenburg and Tsantsabane housing is often informal, on tribal land and rent-free. But Tsantsabane also has a large portion of mine-provided rental housing. This section discusses the implications of mine closure for the main housing types in the cluster analysis.

Several risks exist for Cluster 1 (non-mining, long-time residents with mainly homeownership). This group's non-mining jobs might come under pressure when decline sets in. Those working in electricity generation directly linked to coal might be the first to be affected. But the evidence other studies suggests that businesses and schools may be affected first because lower demand for goods and services may mean business closures and a decline in the number of children in the town (Marais, 2013). This will trigger changes to the staff in education or in other social services. In the Free State Goldfields, when mining declined, the better-skilled people moved first and found jobs more easily (Marais, 2013). But some, especially younger, households and those with mortgage finance might be locked into a location that has lost its economic focus. Although severance packages might be available for some, this group is unlikely to benefit from this kind of relief at scale. The danger is that house prices might drop and selling the house might not raise enough to cover the outstanding mortgage. The housing market might also decline to a point where it makes no sense to sell the house. This group is also the group of people who have developed the most place attachment because they have lived in the area for a long time and have assets.

People in Cluster 2 are extremely vulnerable to mine closure, especially those dependent on mortgage finance. This group of people work directly for the mining companies. They are homeowners with mortgages. They have probably also invested in their house and expect a return on their investment. Many of the people in this group are first-generation middle-class households who have benefited from government policies that link mining, housing and wealth creation. Although their levels of place attachment are lower than Cluster 1's, they are more dependent on continued mining to continue realising their wealth. A rapid decline or closure will disrupt the progress they have made during mining booms and probably lock some of them into their current location if they cannot sell their house.

Group 3 are largely mineworkers who have opted for rental housing. They are to some degree linked to those in Clusters 1 and 2 as they rent houses from homeowners in those two groups. This group's dependence on mining is mainly in terms of their jobs. Should they lose their jobs, they will be mobile to find alternative work or return to a second home. Their future does not depend on the viability of the specific settlement. Their ties are only to their landlords or, in the case of Tsantsabane, to the mining company. Kumba in Tsantsabane is a good example of a mining company protecting its workers from the risks of permanent settlement by providing rental housing.

Table 6.4 Main implications of mine closure for the three housing clusters

Cluster group	*Housing characteristics*	*Mine closure risks*	*Mine closure advantages*
Cluster 1	Ownership Non-mining households – secondary links to the mining economy Homeowners become landlords Housing finance common Formal house Large percentage of respondents unhappy with the house Good access to services	• Considerable decline in the value of housing assets • Inability to sell because there is no market demand (lock-ins) • Skilled people will move first • Owners who rent out their properties might lose their rental income • Might not be able to pay mortgages • Unhappy respondents might be the first people to move out of the area • Closure might hamper the municipality's ability to continue providing services	• Houses might become more affordable for the lower-middle class • A formal house might be a good asset and provide stability in a period of mine closure • Mine decline might reduce levels of unhappiness and the social disruption associated with mining development
Cluster 2	Mineworkers Formal Ownership and rental The lowest percentage of respondents unhappy with the house Best access to services Substantial investments in repairs Lowest preference to leave the area	• Considerable decline in the value of housing assets • High place attachment means mobility will be difficult • Inability to sell because there is no market demand (lock-ins) • Mineworkers unable to find work outside the mining industry	• Renters will be mobile • Those with a second home elsewhere could use it as a diversification strategy • Houses might become more affordable for the lower-middle class • A formal house might provide a good asset and stability in a period of mine closure • Mine decline might reduce levels of unhappiness and the social disruption associated with mining development • Mineworkers will not find jobs outside mining easily
Cluster 3	Mineworkers Largely renters Backyard dwellings Average access to services	• Finding another job might be difficult	• Less backyard dwelling can reduce pressure on existing infrastructure • No local housing assets and lower place attachment will facilitate easy mobility
Cluster 4	Low-income mobile people Informal housing Not directly dependent on mining Lowest level of place attachment	• As the mining economy declines, these households will find it difficult to survive as they are secondary beneficiaries of the mining industry	• Will be highly mobile and relocate easily

Cluster 4 is a group of people in informal housing, often in one-person households and engaged in informal work. In each of the three areas, this group has been the most recent to move in and is likely to be the first to move out when decline sets in. Mobility is their most important advantage. Furthermore, although the evidence from the literature shows that this group might decrease over time, informal housing remains a prominent feature of mining areas in decline.

Table 6.4 summarises the main findings discussed above, showing how people in the various housing types are placed as regards mine closure. For some, their housing type will be advantageous; for others, it could be a liability.

The profiles shown in the cluster analysis also have implications for local governments. Backyard rental requires a different approach to urban services and, in many cases, a change in land use regulations. Encouraging backyard rental could help prevent urban sprawl. Mining municipalities should consider providing rental housing for mineworkers, either family housing as in Tsantsabane or smaller units catering for individuals. This type of housing could be better suited for mining areas, reduce the dependencies that mining creates and allow for more flexible approaches at decline, as the evidence in this section has shown.

The evidence from Tsantsabane points to the value of a public-private partnership in the development of mineworker housing (Stewart & Drewes, 2018). It has enabled the municipality and the mining companies to work together in creating both rental and ownership housing. The backyard approach to rental housing is to make it easy for private households to add rooms or backyard rooms for rent.

6.4 Housing and wealth

6.4.1 Housing values

In this section, I profile various aspects of housing investments and finance in order to determine the risk of mine closure for those with housing assets in the three case study areas (see Table 6.5).

As I cautioned in Chapter 1, the data is not necessarily representative of each case study area, and in some cases, I used very small numbers of responses for the calculations. Table 6.5 points to several conclusions. First, more mining than non-mining households are dependent on mortgage finance. In all three areas, the percentage of mining households with a mortgage was higher than that of non-mining households. The low percentages in Emalahleni result from peculiarities of the sampling and limited access to middle-income households. In Rustenburg and Tsantsabane, the mortgage levels for mining households are similar to the current national average of about 30%. In Tsantsabane's case, a substantial portion of the mortgage data is likely to be leasehold agreements through Assmang and not mortgages. There was no way we could distinguish

Table 6.5 Profile of housing wealth and mortgage dependence

Indicator	*Emalahleni (2017)*		*Rustenburg (2018)*		*Tsantsabane (2015)*	
	Non-mining households	*Mining households*	*Non-mining households*	*Mining households*	*Non-mining households*	*Mining households*
Households who own their house (%)	42	33	40	34	77	51
Households with mortgages/leaseholds (%)	10	12	10	37	13	29
Average mortgage amount outstanding, 2018 ZAR values	Unable to determine	170,458	Unable to determine	335,273	Unable to determine	733,156
Number of respondents who indicated mortgage value		21		17		37
Estimated market value of house, 2018 ZAR values	228,030	1,010,205	630,455	1,138,915	249,645	559,738
Number of respondents who indicated estimated market value	37	55	57	51	63	71
Bought house in 1980 or before (%)	5	6	12	4	5	2
Bought house 1981–1990 (%)	13	10	5	1	11	9
Bought house 1991–2000 (%)	23	26	13	6	27	12
Bought house 2001–2010 (%)	35	33	50	37	30	17
Bought house after 2010 (%)	24	25	20	52	27	60
Repairs done in the past two years (%)	23	23	24	24	29	24
Average amount spent on repairs, 2018 ZAR values	42,549	54,787	71,057	17,745	27,971	31,705
Number of respondents who indicated repairs	31	27	42	55	23	43
Owns another property (%)	4	4	6	15	5	10

Source: Household surveys.

Note: As defined in Chapter 5, a 'mining household' is a household with at least one member employed in mining and a 'non-mining household' is a household that has no member employed in mining.

between the two housing finance systems during the interviews we conducted for our surveys.

Although many did not answer our question about how much they still owed on their mortgage, I was able to roughly calculate the average outstanding amounts for mining households in 2018 ZAR values. The largest amounts owing were in Tsantsabane. Sampling and access issues complicate these results, as does the fact that the housing developments in Tsantsabane are more recent, created by Assmang between 2011 and 2015. Housing developments in Emalahleni and Rustenburg have taken place over a much longer period and so the mortgages are older and the outstanding amounts smaller.

Mining households in Tsantsabane that have housing finance run the highest risk for possible decline or closure. Table 6.5 shows that 84% of the mining households in Tsantsabane had bought their houses since 2010. The corresponding figures in Emalahleni and Rustenburg are 48% and 76%. The big difference in Tsantsabane is that the mining households share the risks with the mining company, as the company still holds the title. The difference between the outstanding housing finance amount and the current value of the house highlights the risk in Tsantsabane: the average amount owing exceeds the estimated average current value of the house. The slump in iron prices in 2015 and more recently has had a detrimental effect on the housing market in Tsantsabane. We do not see these patterns in Emalahleni and Rustenburg.

Table 6.5 also shows that mining households own higher valued properties than non-mining households. Such a house could be the biggest single asset that a household owns. Mine decline and closure could erode its value. It could lead to housing finance problems for those households linked to a mortgage or a leasehold agreement. With closure they might be unable to pay off these loans.

Table 6.5 also shows that mining and non-mining households have second homes. Although the percentages are small, in Rustenburg and Tsantsabane they are higher for mining than non-mining households. In Emalahleni, the percentages for the two are similar. In the Emalahleni survey, 65% of mining households with a second home said it was located outside Emalahleni, and the corresponding figure in Rustenburg was 82%. Most of the Rustenburg respondents' second homes are in the Eastern Cape, although a fair number are in Gauteng. Tsantsabane respondents' second homes were mostly in the Northern Cape.

Although the percentages are small, they are evidence that some mineworkers are creating or maintaining their housing assets elsewhere and not in the town where they work. This phenomenon requires more research with larger samples. The main reasons for these decisions by mining households are probably historical homeownership elsewhere and a deliberate decision not to accumulate housing assets in a place with long-term liabilities like a mining town. In Tsantsabane, mining households with property elsewhere had substantially lower levels of place attachment than those without, as evidenced by the fact that 27% of them had a moderate or strong preference to leave (not shown in the table). The average for this indicator for all mining households in Tsantsabane was 10%. In Emalahleni and Rustenburg, having property elsewhere did not affect the levels of place attachment as the percentages for mining households with and without property elsewhere were the same. The remote location of Tsantsabane is probably one of the main reasons why it differs from the other two areas.

6.4.2 *Household assets of owners and renters*

The section above showed that mineworkers hold substantial housing assets in the three mining towns. Research on wealth and assets in mining areas has found statistically significant differences between mining and non-mining households (Burger & Geldenhuys, 2018; Geldenhuys et al., 2021; Geldenhuys & Burger, 2022). Mining households have higher incomes and more household assets. This section looks at the descriptive statistics for mining households in the three case study areas to see whether the homeowners have more assets than the renters (see Table 6.6).

Table 6.6 shows that on average, the homeowners in all three locations have more household items. The largest difference is in Emalahleni, where homeowners have 68% of the list of household items but renters only 54%. Rustenburg has the lowest percentages, with homeowners having on average 51% and renters

Table 6.6 Asset holding by tenure type for mining households in the three areas, in percentages

Asset	*Emalahleni*		*Rustenburg*		*Tsantsabane*	
	Owners	*Renters*	*Owners*	*Renters*	*Owners*	*Renters*
TV	97	93	87	79	99	95
Radio	85	61	78	79	64	59
DVD/BluRay	84	53	65	62	84	78
Pay TV	90	79	72	68	90	84
Computer	53	41	25	27	50	54
Vacuum cleaner	42	19	4	12	27	43
Washing machine	91	43	45	41	83	69
Dishwasher	8	2	2	9	7	9
Fridge/freezer	98	93	88	79	99	95
Stove/oven	98	94	90	88	98	57
Microwave	89	78	66	62	92	83
Landline phone	8	2	2	3	2	5
Mobile phone	99	100	97	100	98	100
Vehicle	73	49	42	47	63	73
Motorcycle	2	2	4	0	1	1
Average for household items	67.8	53.9	51.1	50.4	63.8	60.3
Internet	95	91	67	56	84	87

Source: Household surveys.

50%. The renters who have the most items are in Tsantsabane, with 60%. The corresponding figures for Emalahleni and Rustenburg renters are 54% and 50%.

We can draw two main conclusions from the above data. First, although homeowners have more household items than renters, the difference is fairly small – less than 1% in Rustenburg and Tsantsabane. Second, the high percentages of items owned by those who rent in Tsantsabane are noteworthy, since it is here that mining companies have provided rental housing, unlike companies in the other two areas. This shows that households in rental housing can accumulate assets. The Tsantsabane renters' average percentage is even higher than the Rustenburg homeowners' average. This indicates that company-provided rental housing *can* provide stability and enable asset creation, and that asset creation is not necessarily dependent on homeownership.

6.4.3 *Self-ranking of wealth*

Here, we look again at the self-ranking of wealth exercise that was discussed in Chapter 5, to see whether mining household owners and renters assessed their level of wealth differently.

Figure 6.1 shows that Emalahleni and Rustenburg owners ranked their wealth slightly higher than the renters. Emalahleni owners ranked their wealth at age 15 lower than the renters did, but consistently higher since then. The difference is fairly small and has remained much the same over the period. Rustenburg owners ranked their wealth at age 15 very slightly higher than the renters and this very

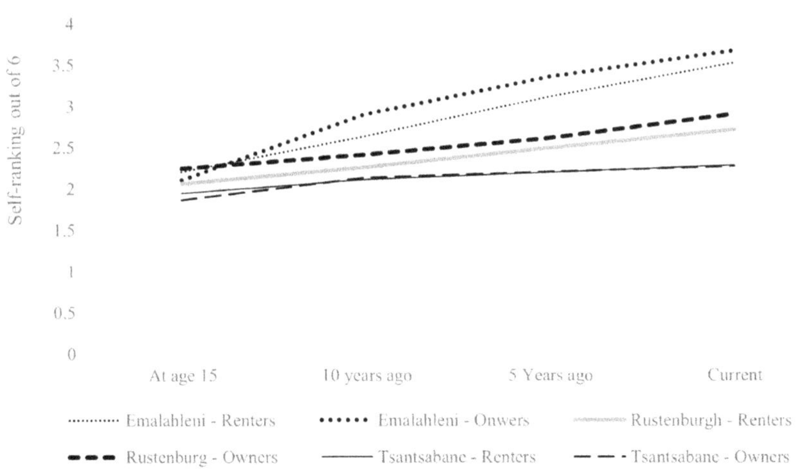

Figure 6.1 Self-ranking of wealth by renters and owners in the three case studies.

Source: Household surveys.

Notes: 'Current' means at the time of the survey: 2015 for Tsantsabane, 2017 for Emalahleni and 2018 for Rustenburg. The Tsantsabane survey did not include a ten-year ranking. I used the midpoint between the five year and aged 15-year rankings.

small difference has remained the same over the period. In Tsantsabane, where Kumba provided rental housing, there is a very small difference in ranking of wealth at age 15, but no visible difference since them.

The inferences are clear here. People who own houses in the three areas do not rank their wealth very much higher, or see a larger increase over the years, than people who rent. In Tsantsabane, which differs from the other two areas in having mine-provided rental housing, we see virtually no difference between the owners' and the renters' rankings. The figures do not support the policy assumption that homeownership makes people wealthier. Further, the example of Tsantsabane shows that decent rental housing provided by a mining company can eliminate the difference between owners' and renters' wealth.

6.5 Profiles of renter mining households compared with owners

In this section, I compare the profile of mining households that rent with the profile of those that own (Table 6.7), to see whether renters' demographic and

Table 6.7 Profiles of mining household renters compared with owners

Indicator	*Emalahleni*		*Rustenburg*		*Tsantsabane*	
	Renting	*Owning*	*Renting*	*Owning*	*Renting*	*Owning*
Percentage of households	41	59	58	42	43	57
Household size	1.31	2.67	2.61	2.93	2.54	3.30
Households with someone living elsewhere ten years ago (%)	50	14	46	26	70	35
Living in informal house (%)	4	4	9	15	8	28
Living in a backyard (%)	44	16	22	6	7	4
Number of rooms	2.80	6.40	3.33	4.63	5.11	4.84
Number of rooms per person	2.14	2.40	1.25	1.58	2.17	1.46
Average rent paid, 2018 ZAR values	1418	n.a.	2281	n.a.	1,284	n.a.
Average income, 2018 ZAR values	18,449	n.a.	14,661	n.a.	13,882	n.a.
Average rent per room, 2018 ZAR values	506	n.a.	684	n.a.	250	n.a.
% of income spent on rent	8	n.a.	16	n.a.	9	n.a.
Preference for ownership (%)	47	n.a.	54	n.a.	59	n.a.
Sanitation shared (%)	69	40	53	28	35	25
Electricity cuts because of non-payment (%)	39	18	36	0	9	14
Households unhappy with house (%)	18	5	31	37	14	26

Source: Household surveys.

household-related features might differ from those of the owners and thus explain some reasons for renting. I made a distinction between formal and informal housing.

Table 6.7 shows that between 40% and 60% of mining households in the three areas are renters: 41% in Tsantsabane, 43% in Emalahleni and 58% in Rustenburg. In Chapter 5, I explained that Rustenburg has the largest percentage of migrant workers and a long history of migrant labour. On the evidence above, the continuation of migrant work appears to favour rental housing.

The table shows demographic differences between renters and owners. In all three locations, the renter households are smaller than the owner households. Smaller household sizes can be related to migrant labour, but they can also point to a specific demographic sector (often one-person households) needing rental housing. In Emalahleni and Rustenburg, the owners have larger houses than renters (judging by the number of rooms). In Emalahleni, the difference is stark, with owners having an average of 6.4 rooms and renters only 2.8. The reverse is true of Tsantsabane: renters have larger houses than owners. The situation is the same for the density of occupation: in Emalahleni and Rustenburg, the owners have more rooms per person than renters, and but the reverse is true of Tsantsabane. Again, the figures for Tsantsabane emphasise the importance of company-provided rental housing.

The average rents paid are highest in Rustenburg, at R2281 per month, compared to the R1418 in Rustenburg and R1284 in Tsantsabane. The average rent per room is highest in Rustenburg and these renters pay the largest percentage of income on rent (16%). The lowest monthly rent per room is in Tsantsabane (ZAR 250) and renters in Emalahleni pay the smallest percentage of income on rent (8%). This low percentage is probably because Emalahleni has the highest household income.

The percentages of informal housing are small, particularly for renters. Many renters in Emalahleni and Rustenburg are in backyards paying rent to the homeowners. This housing type also explains the smaller household size of the renters. Backyard renting is much less common in Tsantsabane, where the mine provides rental housing. This has reduced the local market for backyard rentals (although some backyard rentals are available in Tsantsabane). The table shows one major problem for backyard dwellers: many more of them have to share toilet facilities.

6.6 Crime and social disruption

Chapter 5 provided a detailed overview of factors contributing to experiences of social disruption. I investigated whether mining and non-mining households that rent or own have different perceptions of crime rates (see Tables 6.8 and 6.9). Here, I look again at the evidence from respondent perceptions about a lost wallet and the level of crime.

Table 6.8 Likelihood that a wallet would be returned, in percentages

Not likely at all that it would be returned by…	*Emalahleni*		*Rustenburg*		*Tsantsabane*	
	SCB	*Stranger*	SCB	*Stranger*	SCB	*Stranger*
Total population	73.3	90.5	86.3	94.9	78.0	92.2
Non-mining households – renters	88.8	94.1	88.7	96.4	82.3	95.2
Mining households – renters	71.3	89.1	86.9	92.6	80.0	92.2
Non-mining households – owners	82.8	97.3	85.2	95.7	80.7	92.0
Mining households – owners	55.4	83.7	82.5	96.3	70.1	90.2

Source: Household surveys.
Note: SCB = someone close by.

Table 6.9 Average crime index scores

	Emalahleni	*Rustenburg*	*Tsantsabane*
Total population	20.1	20.7	20.1
Non-mining households – renters	20.4	20.6	18.6
Mining households – renters	19.1	18.3	16.7
Non-mining households – owners	23.0	22.6	21.8
Mining households – owners	18.9	20.1	20.1

Source: Household surveys.
Note: To assess people's experience of crime in the three study areas, the survey asked respondents to rate their perceptions of the levels of six types of crime in their area: burglaries, muggings or thefts; violence between members of the same household; violence between members of different households; gangsterism; murder, shootings or stabbings; and drug or alcohol abuse. A crime index was developed by rating the responses on a scale of 1 to 5, with 4 being 'fairly common' and 5 'very common'. The average was then calculated by summing the responses for each of the six types of crimes.

Both tables show that, on average, renters, and particularly mineworkers who rent, have lower perceptions of crime than owners. Once again, the renters in Tsantsabane seem to be different. Their perception of crime is far lower than any other group's. I think this is directly related to the major investment by Kumba in rental housing for mineworkers. The higher level of crime experienced by non-mining households may be because these are existing residents experiencing social disruption because of mining. Long-time residents find the influx of mineworkers and a change in the local culture disruptive and often do not trust newcomers.

The most important conclusion from the tables is that non-mining households may be experiencing more crime and social disruption because of mining than the mining households. Should decline set in, mining households might experience it as much as non-mining households, and maybe even more.

6.7 Conclusion

This chapter first discussed the Africa Mining Vision, the African Union's attempt to ensure that the continent benefits from mining. The Africa Mining Vision is largely one-directional: it aims to achieve modernity for the African population through mining. It does not consider the dependency created by the mining industry or the disruptive reality of mine closure. Its argument is simple: mining must create benefits and help establish modernity. Absent from its thinking is the threat of decline shattering the dream of modernity and putting African mining cities at risk. The African literature on mine closure provides a bleak picture of mine closure realities in Africa. Local governments and households do not recognise the dependencies that mining creates. High expectations of wealth and modernity are often accompanied by poor planning.

The chapter profiled the housing types in the three case study areas. Government policies focusing on housing, ownership and wealth appear to have worked well. For example, formal housing has outstripped informal housing in some cases. There is evidence of mining creating assets. But households that have benefited from the policy are now at risk in a case of mine decline or closure. Those who own their house might lose it because they cannot pay the mortgage or find their house has lost value because of market decline. Place attachment and stability might not be enough and could become an obstacle to finding alternative employment or housing.

Those households in informal or rental housing are likely to manage decline and closure better. They can move elsewhere, ending their rental contracts or simply leaving their informal housing. Their migration histories and their lower levels of place attachment will benefit them when they want to relocate or consolidate a second home. The chapter showed in particular the value of the rental housing developed by Kumba in Tsantsabane. Owning a house is not essential for home comforts and asset creation. Rental housing can also provide adequate housing for mineworkers and enable them to accumulate household assets. It offers flexibility in a town whose economic stability is temporary and unlikely to provide long-term prosperity, as is the case with most mining towns.

References

African Union, 2009. *The Africa Mining Vision: Africa Mining Vision African Minerals Governance Framework*. Addis Ababa: African Union.

Béland, D., Campbell, B., Coderre, M. & Haang'andu, P., 2021. Policy change and paradigm shifts in Sub-Saharan Africa: implementing the Africa Mining Vision. *Canadian Journal of African Studies*. https://doi.org/10.1080/00083968.2021.1886955.

Bowman, A., Frederiksen, T., Bryceson, D., Childs, J., Gilberthorpe, E. & Newman, S., 2021. Mining in Africa after the supercycle: new directions and geographies. *Area*. https://doi.org/10.1111/area.12723.
Bradbury, J. & St-Martin, I., 1983. Winding-down in a Quebec mining town: a case study of Schefferville. *Canadian Geographer*, 27(2), pp. 128–144.
Bryceson, D. & MacKinnon, D., 2013. Eureka and beyond: mining's impact on African urbanisation. *Journal of Contemporary African Studies*, 30(4), pp. 513–537.
Burger, P. & Geldenhuys, J., 2018. Work, wages and welfare in Postmasburg. In: L. Marais, P. Burger & D. Van Rooyen, eds, *Mining and Community in South Africa: From Small Town to Iron Town*. London: Routledge, pp. 173–198.
Busia, K. & Akong, C., 2017. The African mining vision: perspectives on mineral resource development in Africa. *Journal of Sustainable Development Law and Policy*, 8(1), pp. 145–192.
Cawood, F., 2017. Policy elements for twenty-first century African mining: strengthening the existing African Mining Vision themes with cross-cutting subsets. *Journal of Science and Technology Policy Management*, 9(1), pp. 87–101.
Cloete, J. & Marais, L., 2021. Mine housing in the South African coalfields: the unforeseen consequences of post-apartheid policy. *Housing Studies*. https://doi.org/10.1080/02673037.2020.1769038.
Dauda, S., 2020. Operationalising the 'Africa Mining Vision': critical reflections from Ghana. *Canadian Journal of Development Studies*, 41(3), pp. 504–524.
Ferguson, J., 1999. *Expectations of Modernity: Myths and Meanings of Urban Life on the Zambian Copperbelt*. Berkeley: University of California Press.
Filitz, J., 2011. *Mining for Development? A Socio-ecological Study on the Witbank Coalfields*. Durban: Unpublished Master's dissertation, University of KwaZulu-Natal.
Frazer, A., 2010. Introduction: boom and bust on the Zambian Copper belt. In: A. Fraser & M. Larmer, eds. *Zambia, Mining and Neoliberalism: Boom and Bust on the Globalised Copperbelt*. New York: Palgrave Macmillan, pp. 1–30.
Geldenhuys, J. & Burger, P., 2022. Household welfare in Emalahleni. In: L. Marais, P. Burger, M. Campbell, S. Denoon-Stevens & D. van Rooyen, eds. *Coal and Energy in South Africa: Considering a Just Transition*. Edinburgh: Edinburgh University Press, pp. 44–60.
Geldenhuys, J., Pool, A. & Burger, P., 2021. Wages and welfare in Rustenburg. In: L. Marais, M. Campbell, S. Denoon-Stevens & D. Van Rooyen, eds. *Mining and Community in the South African Platinum Belt: A Decade after Marikana*. New York: Nova, pp. 93–118.
Gewald, J. & Soeters, S., 2010. African miners and shape-shifting capital flight: the case of Luanshya/Baluba. In: A. Fraser & M. Larmer, eds. *Zambia, Mining and Neoliberalism: Boom and Bust on the Globalised Copperbelt*. New York: Palgrave Macmillan, pp. 155–183.
Gough, K. & Yankson, P., 2012. Exploring the connections: mining and urbanisation in Ghana. *Journal of Contemporary African Studies*, 30(4), pp. 651–668.
Gough, K., Yankson, P. & Esson, J., 2019. Migration, housing and attachment in urban gold mining settlements. *Urban Studies*, 56(13), pp. 2670–2587.
Hilson, G., 2020. The Africa Mining Vision: a manifesto for more inclusive extractive industry-led development? *Canadian Journal of Development Studies*, 41(3), pp. 417–431.
Jønsson, J.B.D., Kinaboc, K. & Shand, M., 2019. Getting grounded? Miners' migration, housing and urban settlement in Tanzania, 1980–2012. *The Extractive Industries and Society*, 6(3), pp. 948–959.

Jønsson, J. & Bryceson, D., 2017. Beyond the artisanal mining site: migration, housing capital accumulation and indirect urbanisation in East Africa. *Journal of Eastern African Studies*, 11(1), pp. 3–23.

Kamete, Y., 2012. Of prosperity, ghost towns and havens: mining and urbanisation in Zimbabwe. *Journal of Contemporary African Studies*, 30(4), pp. 589–609.

Kamlonger, P., 2020. Domesticating the Africa Mining Vision in Malawi: approaches and experiences. *Canadian Journal of Development Studies*, 41(3), pp. 467–485.

Knierzinger, J., 2017. Mining towns as portals of globalisation: the arrival of the global aluminium industry in West Africa. *Comparativ*, 27(3–4), pp. 94–110.

Knierzinger, J., 2018. *Bauxite Mining in Africa: Sociopolitical Implications of Transnational Corporate Governance*. London: Palgrave Macmillan.

Knierzinger, J. & Sopelle, I., 2019. Mine closure from below: transformative movements in two shrinking West African mining towns. *Extractive Industries and Society*, 6, pp. 145–154.

Larmer, M., 2006. 'The hour has come at the pit': The Mineworkers' Union of Zambia and the Movement for Multi-Party Democracy, 1982–1991. *Journal of Southern African Studies*, 3(2), pp. 293–312.

Lee, C., 2014. *The Specter of Global China. Politics, Labor, and Foreign Investment in Africa*. Chicago, IL: University of Chicago Press.

Littlewood, D., 2014. 'Cursed' communities? Corporate Social Responsibility (CSR), Company Towns and the Mining Industry in Namibia. *Journal of Business Ethics*, 120(1), pp. 39–63.

Marais, L., 2013 Mine downscaling in the Free State Goldfields. *Urban Forum*, 24, pp. 503–521.

Marais, L. & De Lange, A., 2021. Anticipating and planning for mine closure in South Africa. *Futures*, 125, p. 102669.

Marais, L., Ntema, J., Cloete, J. & Lenka, M., 2021. Mine housing, assets and informality in Rustenburg: Implications of mine closure. In: L. Marais, M. Campbell, S. Denoon-Stevens & D. Van Rooyen, eds. *Mining and Community in the South African Platinum Belt. A Decade after Marikana*. New York: Nova, pp. 141–158.

Mfune, O., Chansa-Kabali, T., Chisola, C. & Manchisi, J., 2020. The legacy of mine closure in Kabwe, Zambia: what can resilience thinking offer to the mining sustainability discourse? In: G. Fasparatos, M. Naidoo, A. Ahmed, A. Karanja & K. Fukushi eds. *Sustainability Challenges in Sub-Saharan Africa: Insights from Eastern and South Africa*. Singapore: Springer Nature, pp. 167–188.

Ntema, J., 2019. Rustenburg: boom and bust in a mining town. In: L. Marais & V. Nel, eds. *Space and Planning in Secondary Cities: Reflections from South Africa*. Stellenbosch: Sun Media, pp. 203–218.

Nyame, F., Grant, A. & Yakovlevac, N., 2009. Perspectives on migration patterns in Ghana's mining industry. *Resources Policy*, 34(1–2), pp. 6–11.

Owen, J., Kemp, D. & Marais, L., 2021. The cost of mining benefits: localising the resource curse hypothesis. *Resources Policy*, 74, pp. 102239.

Rubbers, B., 2010. Changing workers' rights in the Democratic Republic of the Congo. *Review of African Political Economy*, 37(125), pp. 329–344.

Stats SA, 1996. *Census Data 1996*. Pretoria: StatsSA.

Stats SA, 2001. *Census Data 2001*. Pretoria: StatsSA.

Stats SA, 2016. *Community Survey 2016*. Pretoria: Stats SA.

Stewart, T. & Drewes, E., 2018. The Khumani approach to homeownership in Postmasburg. In: L. Marais, P. Burger & D. Van Rooyen, eds. *Mining and Community in South Africa: From Small Town to Iron Town*. London: Routledge, pp. 157–170.

Sturman, K., Toledano, P., Akayuli, C. & Gondwe, M., 2020. African mining and the SDGs: from vision to reality. In: M. Ramutsindela & D. Mickler, eds. *Africa and the Sustainable Development Goals. Sustainable Development Goals Series*. Cham: Springer, pp. 59–69.

Van der Watt, P. & Marais, L., 2019. Normalising mining company towns in Emalahleni, South Africa. *The Extractive Industries and Society*, 6(4), pp. 1205–1214.

7 Mining and informal housing

7.1 Introduction

Many African mining towns have a dual settlement system, formal and informal, with informality predominating. Informal housing springs up in mining towns for various reasons. Most obviously, governments struggle to provide housing for the influx of migrants seeking work. The uncertainty and temporariness of mining is another notable cause. If the economic future is doubtful, people and governments do not want to invest in houses and infrastructure. Living with impermanence and being mobile has become the norm in many mining communities. The case studies on which this book is based covered only commercial large-scale mining, but the findings have relevance to artisanal mining too. Low and volatile incomes associated with artisanal mining in many African countries further contribute to an upsurge in informal living.

In Chapter 5, I noted that migrant labour in South Africa continues despite various attempts by the government to stop it. Mobility and migration are livelihood diversification strategies. However, informal housing is one of the consequences of migration and mobility. In Chapter 6, I referred to the high percentages of informal housing in the three case study areas and showed that mining municipalities experiencing mining decline have high levels of informal housing (although not as high as those where mining is growing). These high levels of informality are evident despite the South African policy aims of formality, homeownership, stability and local development through mining (see Chapter 4). I have explained how these government policies are both path and goal dependent. However, policy does not take sufficient cognisance of the way mine closure disrupts the policy aims.

In this chapter, I review the African and South African literature and use the material from the three case studies to analyse the question of informal housing. Although local planning attempts to counter the growth of informal housing, many mineworkers choose to live in informal housing irrespective of the efforts of mining companies and governments. In this chapter, I define informality as 'those activities that fall outside government regulation, control, support and taxation' (Stark & Teppo, 2021). Informal housing is, of course, not an isolated phenomenon but part of the informal economy that develops around the mining

DOI: 10.4324/9781003262015-7

industry, including informal trading, informal renting of houses and sometimes illegal activities such as prostitution and drug dealing. Setting aside structural reasons like poverty that contribute to informal housing, I argue that informal housing in mining areas is a rational response to the temporariness and uncertainty of mining. It is a way to manage the likelihood of mine decline and avoid long-term dependency.

7.2 The literature

7.2.1 Mining and informal housing in Africa

We still have much to learn about informal housing and its relationship with mining (Mwango, 2006). Lindell and Utas (2012, p. 409) say that 'What appears to outsiders as chaotic informal activity is usually structured by webs of relationships that lend life in highly uncertain settings a degree of predictability'. Informal housing is a rational response to minimise uncertainty. In Africa, artisanal mining is a major contributor to informality and informal housing (Hamann, 2004; Hilson, 2013; Van Bockstael, 2014). I highlight seven findings from the literature on informal housing in mining areas in Africa, using the following sources: Gough and Yankson (2012), Grätz (2013), Littlewood (2014), Steyn and Atamelang (2014), Jønsson and Bryceson (2017), Gough et al. (2019), Jønsson et al. (2019), Bryceson et al. (2020), Amofah et al. (2021) and Rubbers (2021).

First, informal housing has been a common feature of urban living in many African cities since World War II. Increased urbanisation in the primary cities has been the main reason for the rapid increase. The responses to urbanisation in African cities in the 1950s were threefold: investment in public rental housing, demolition of informal housing and policies discouraging urbanisation. None of these strategies succeeded in curbing urbanisation and the growth of informal settlements. In the 1970s and 1980s, the response slowly changed to informal settlement upgrading, with many projects being funded by the World Bank. Yet this upgrading has not reduced the number of people in these settlements and many cities still have large numbers of informal houses. However, the informal urbanising associated with mining differs from the urbanising after World War II. Mining-induced urbanisation is usually in secondary cities and the smaller towns rather than the primary cities. These secondary cities and small towns have limited institutional capacity to manage urbanisation. Many African countries do not have decentralised local governments. Where they do exist, their roles, functions and financing are inadequate. Although circular migration also happens in the primary cities, the urbanisation in those cities is usually more permanent than in secondary cities.

Second, mineworkers and other migrants use informal housing to deal with uncertainty. Living in informal housing near the mines helps people manage the temporary nature of mining. Temporariness and informal housing go hand in hand. Informality enables people to stay mobile by not accumulating large housing assets in a mining area (see Chapter 5).

Third, contradicting the strategy of accepting temporariness and adapting to it, policymakers emphasise the need for stability and the development of place attachment in mining towns. They think that mining growth will eventually help address the informal housing problem and that appropriate planning should support this long-term ideal. There are examples from Ghana where mine closure contributed extensively to the development of self-build luxury houses because of the severance packages paid by the mining companies (Amofah et al., 2021). Place attachment does develop in some places and housing investments could support the development of place attachment.

Fourth, research from Tanzania shows that informality in mining towns can result from households deciding to invest in housing elsewhere. Jønsson and Bryceson (2017, p. 3) say informal housing is a way to 'make strategic house building investments in larger towns and cities in anticipation of mine decline'. Informal housing in mining towns is a rational response by mineworkers. In Namibia, companies have acknowledged that some mineworkers do not want houses near the mining operations. Littlewood (2014, pp. 49–50) says mineworkers in Namibia see mining settlements as places 'to live in the short- and medium-term while they work, but do not intend to stay any longer'; they are 'economic migrants' who 'will leave upon mine closure in search of new opportunities'. Littlewood's study shows that the mines provide decent housing for their mineworkers near the mines while the mineworkers invest their extra savings in housing in places of their choice. However, it is not only mineworkers who migrate and settle informally. Traders and service providers follow the mineworkers.

Fifth, like migration and mobility, informal housing can be a strategy to maintain flexibility. Several family members remain behind in a large house in a town, while those who are mineworkers migrate to new mines (often artisanal). The family members in the town can engage in other economic activities. This means that the settlements near the mines are often informal as the miners need to be able to keep moving and do not want to invest near the mines. Bryceson et al. (2020, p. 446) note that 'mining site residents highly mobile lives entail toleration of temporary, inadequate housing in infrastructurally deficient, polluted and unsafe mining environments, a situation at odds with their aims for lifestyle'.

Sixth, Bryceson et al. (2020) point to local governments' inability to manage the influx of people into mining areas. In the absence of appropriate responses by local governments, mining companies have started to actively support worker camps. These camps affect the way mining company employees interact with or are buffered from local communities. Worker camps are the outcome of fly-in-fly-out policies (see Chapter 5).

Seventh, the research links informal settlements with crime and poor health. Hamann and Kapelus (2004, p. 88) point to the 'growing informal settlements (squatter camps) around the mines, whose residents suffer deteriorating social conditions, including lacking services, crime, and disease'. This view of informal housing is not new and not without merit. But it could lead to an overemphasis on the adverse aspects and distract people from seeing the functional role of informal housing.

The high percentage of informal housing and continued migration has housing policy implications for local governments and mining companies. The transitional nature of mine housing in many mining areas requires acknowledgement in the policy. Although much informal housing develops for structural reasons like poverty, poor planning and lack of infrastructure, for many mineworkers it is a rational response to the temporariness of mining.

Policies should consider both permanence and impermanence in their responses to the growth of mining cities. Context-specific factors should influence where to place the emphasis.

7.2.2 *Mining and informal housing in South Africa*

There is a growing body of research on informal housing associated with mining in South Africa. This section is based on the following sources: Crush and James (1991), Hamann (2004), Hamann and Kapelus (2004), Marais and Venter (2006), Hamann (2009), Nkomo (2014), Bezuidenhout (2015), Bezuidenhout and Buhlungu (2015), Rubin and Harrison (2016), Naumann and Greiner (2017), Marais et al. (2018), Stewart and Drewes (2018), Van der Watt and Marais (2019), Mujere (2020), Marais et al. (2020), Marais et al. (2021) and Marais et al. (2022). In Chapter 4, I explained that the dismantling of the mineworker compounds had mixed consequences, one of them being an increase in informal housing. Four factors contributed to this unexpected outcome.

First, giving mineworkers a choice meant that some did opt for informal housing (see Chapters 4 and 6). In the mid-1990s, Crush and James (1991) predicted that the depopulation of the compound would be a slow process. But within about 20 years, the government has dismantled most of the compounds. Yet the dismantling has not necessarily created better housing for all mineworkers. Many have used their new autonomy and the living-out allowance to live in informal housing.

Second, because mining creates large-scale in-migration, local authorities have been unable to provide enough formal housing for all. The result has been urban sprawl in mining towns in South Africa. Many do not have the financial means to make new land developments and provide bulk infrastructure. Dealing with the influx is difficult and requires partnerships with the mining companies. Tsantsabane is a good example of such a partnership between the local authority and mining companies.

Third, informal housing results from mining companies withdrawing from providing housing. In Emalahleni, for example, the closure of mining villages fuelled an in-migration to the town. The lower levels of informal housing in Tsantsabane (see Chapter 7, Section 7.4) are the direct result of the mining companies' active involvement in mine housing.

Fourth, in Rustenburg, tribal land has contributed to informal housing. Tribal land has limited land use regulations and settlements can occur ad hoc. As many of Rustenburg's mines are on tribal land, many mineworkers and people dependent on the secondary mining economy build their houses near the mines. New mining areas in the Limpopo province have the same problem with informality in tribal areas.

7.3 Informal housing in towns with mining growth and decline in South Africa

We now look at informal housing, household numbers and services levels in municipalities where mining is growing and municipalities where mining is in decline. As in Chapter 5, the mining growth municipalities are seven where mining's share of the local economy was three times more than mining's share of the national economy and the mining economy was stable or growing between 1996 and 2016: Ba-Phalaborwa (copper), Emalahleni (coal), Fetakgomo Tubatse (platinum), Lephalale (coal), Madibeng (platinum), Metsimaholo (coal), Rustenburg (platinum) and Steve Tshwete (coal), and the mining decline municipalities are five that experienced a decline in mining's share in the local economy of at least 15% between 1996 and 2016: Govan Mbeki (coal), Matjhabeng (gold), Matlosana (gold), Merafong City (gold) and Rand West (gold).

Table 7.1 Housing and access to services in mining municipalities in South Africa, 1996–2016

Indicators	*Mining growth municipalities*		*Mining decline municipalities*	
	1996	*2016*	*1996*	*2016*
Households	370,424	969,649	327,154	577,370
Annual household (HH) growth	4.9%		2.9%	
HHs in informal housing	90,235	215,344	126,156	96,726
Annual growth of HHs in informal housing	4.5%		−1.3%	
Percentage of HHs in informal housing	24.3%	22.2%	38.6%	16.8%
HHs with indoor water	141,139	307,536	175,371	301,479
Annual growth of HHs with indoor water	4.0%		2.8%	
Percentage of HHs with indoor water	38.1%	31.7%	53.6%	52.2%
HHs with a flush toilet	157,797	497,434	219,896	511,763
Annual growth of HHs with a flush toilet	5.9%		4.3%	
Percentage of HHs with a flush toilet	42.5%	51.3	67.2%	88.6%

Source: Stats SA (1996, 2016).

Table 7.1 shows that in the municipalities where mining has been growing, household growth outstripped the growth in informal housing. The number of households grew by nearly 5% per annum compared to 4.5% for informal houses. We can see that formal housing provision grew in addition to the growth in informal housing, since the table shows that informal housing decreased as a percentage of total houses, from 24% to 22%. These municipalities made progress with infrastructure access, but large backlogs remain. For example, households with a flush toilet increased from 43% to 51%, but despite this increase, one in two households still does not have waterborne sanitation.

The difference between mining growth and mining decline municipalities remains stark. In the mining decline municipalities, the share of informal housing decreased from 39% in 1996 to 17% in 2016, indicating the outflow of people. The percentage is, however, still higher than the average percentage for South Africa at about 13.5% in 2016.

The percentage of households with waterborne sanitation increased rapidly in the mining decline municipalities. This increase points to the value of decline – increased service levels and a decrease in informal housing. It confirms the importance of informality in mining areas. It ensures that people are mobile and do not become stuck in a declining economy.

Figure 7.1 compares the five mining growth and seven mining decline municipalities by plotting their informal housing growth and household growth.

Mine decline has resulted in a decline in informal housing in three of the five mining decline municipalities, yet all five have seen increases in the number of households. New household formation has occurred despite some of these municipalities reporting negative or very low population growth. There are two main reasons for this. First, this is a general trend in South Africa. Often, it results from

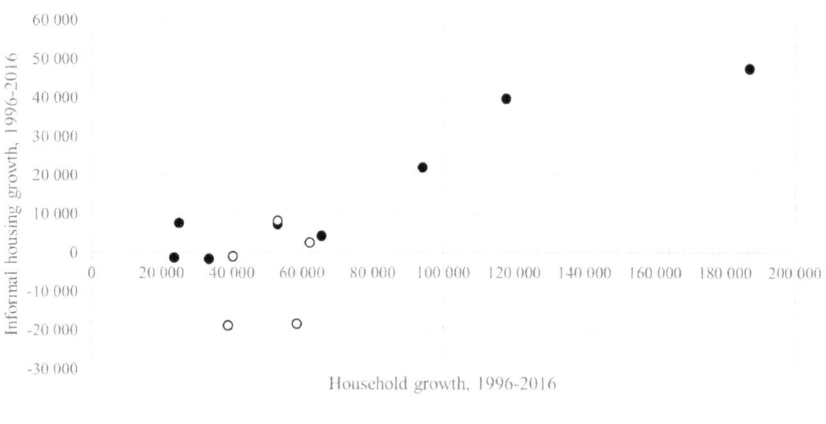

Figure 7.1 Growth of informal housing and households in mining municipalities, 1996–2016.

Source: Stats SA (1996, 2016).

a housing policy that obliges people to form new households to access the housing subsidy programme. Sometimes, new households form in mining areas because of policies promoting sustainable settlements and stability. Providing new stands is part of the policy in these mining cities.

The three municipalities with the largest informal housing and household growth are mining growth municipalities. This confirms the conclusions in earlier chapters. Furthermore, four of the mining growth municipalities experienced low informal housing growth (two even experienced a small decrease) and low growth in the number of households.

7.4 Mining and informality in the case study areas

Informal housing is far more prominent in mining than non-mining towns in South Africa. In this section, I analyse informality in the case study areas.

7.4.1 *Changing patterns of informality*

Figure 7.2 shows changes in the percentages of people in informal housing in the three case study areas and in South Africa from 1996 to 2016.

The percentages of informal houses in the three case study areas were higher than the average for South Africa, apart from those for Tsantsabane in 1996 and 2001. The high percentages are the result of influx in response to mining booms. The effect of Tsantsabane's 2010 boom can be seen in the rapid increase between 2001 and 2011. The slump in iron ore prices in 2015/2016 explains the decrease in 2016. Rustenburg shows a different picture. The percentage of informal houses was very high in 1996 and 2001 (at the height of the platinum boom) and then

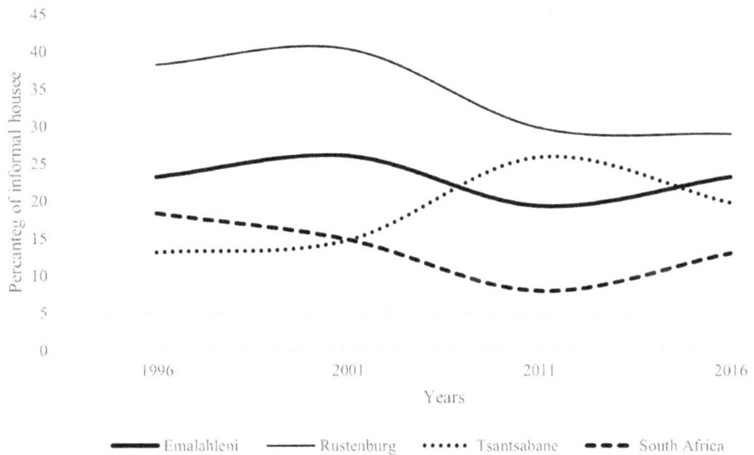

Figure 7.2 Percentage of informal houses in the case study areas, 1996–2016.
Sources: Stats SA (1996, 2001, 2011, 2016).

dropped. But despite the drop in the percentage of informal houses, the real numbers grew. In 1996, 29,000 houses were informal (38% of total housing stock) compared to 76,000 in 2016 (29%). In Emalahleni, the percentage of informal houses in 1996 and 2016 was the same. But in line with the boom since the mid-1990s, there was an increase in the percentage between 1996 and 2001. The real numbers of informal houses increased from 13,000 in 1996 to 35,000 in 2016.

7.4.2 Characteristics of residents in informal housing

Table 7.2 compares the profiles of formal and informal housing households in the case study areas.

We see social stratification based on formal or informal housing in all areas. Households in informal houses have much lower levels of indoor water, sanitation and electricity than those in formal houses. For example, 91% of households in formal houses in Emalahleni have access to waterborne sanitation compared to 13% in informal houses. The employment profiles provide further evidence of the social stratification. In all three areas, substantially more households in informal than in formal housing have no household member formally employed. For example, 54% of households in informal housing in Rustenburg have no one formally employed compared with 34% in formal houses.

Social stratification is also evident in the number of rooms per house. Formal houses are substantially larger than informal houses, particularly in Tsantsabane, where formal houses have an average of 5.5 rooms and informal houses 2.2. The result is more crowding in informal houses: 1.5 rooms per person in formal housing and 0.7 in informal houses. The corresponding figures for Emalahleni and Rustenburg are 1.9 and 1.6 rooms per person in formal houses and 1.1 and 1.5 in informal houses. The smaller informal houses in Tsantsabane are the result of the recent boom. The booms in the other two case studies were less recent and thus there has been more time to increase add-on rooms. Household sizes differ between formal and informal houses. In all three areas, households in informal houses are on average smaller than in formal houses: 1.7 persons per household compared with 4.5 in Emalahleni, 1.8 compared with 4.2 in Rustenburg and 3.3 compared with 5.5 in Tsantsabane. Household size and time spent at a specific place are prominent factors in determining the size of houses.

Only a small percentage of respondents in formal houses are dissatisfied or unhappy with their houses: 33% in Rustenburg and less than 20% in Emalahleni and Tsantsabane. In contrast, more than 50% of the respondents in informal houses in all three areas are dissatisfied or unhappy with their houses.

Although fewer households with someone employed on the mines live in informal housing, the percentages are still fairly high: 16% in Emalahleni, 26% in Rustenburg and 32% in Tsantsabane. The higher percentage for Tsantsabane shows that despite commendable efforts by the mining companies to provide good housing for their employees, not all have been reached by these efforts, and not all may want formal housing.

Table 7.2 Comparison of households in formal and informal housing

Indicators	*Emalahleni*		*Rustenburg*		*Tsantsabane*	
	Formal	*Informal*	*Formal*	*Informal*	*Formal*	*Informal*
Percentage of total housing in sample	85.4	14.6	82.1	17.9	73.8	26.2
Average household (HH) size	2.4	1.7	2.6	1.8	3.6	3.0
Percentage younger than 40	75.7	76.8	65.1	61.1	85.8	86.9
Percentage of one-person HHs	41.4	60.7	32.9	57.6	18.5	34.6
Percentage with someone in HH employed in mining	51.3	16.2	43.7	25.8	41.2	31.9
Percentage of HHs with no person formally employed	14.8	42.2	34.3	54.1	17.8	26.5
Percentage of HHs with self-employment	12.8	14.8	14.8	12.4	7.9	5.8
Had someone in HH migrating in the past ten years	31.3	47.4	32.4	46.4	38.1	49.6
Average number of rooms	4.5	1.9	4.2	1.2	5.5	2.2
Average number of rooms per person	1.9	1.1	1.6	1.5	1.5	0.7
Percentage sharing a toilet facility	51.4	68.1	37.5	81.6	22.6	34.0
Percentage with flush toilet	91.3	13.2	78.5	22.9	97.7	25.2
Percentage with water on the stand	94.7	23.7	98.4	36.5	95.6	40.0
Percentage with electricity	94.8	15.6	98.8	35.6	93.6	30.1
Percentage unhappy with the house	16.4	70.0	33.0	66.5	18.1	53.5
Percentage did repairs in the past two years	25.3	10.1	16.0	7.6	27.9	23.7
Percentage with a house elsewhere	4.3	3.2	9.9	4.6	10.0	2.8

Source: Household surveys

Households in informal housing are more likely to have someone who migrated to the area in the past ten years: 47% in Emalahleni, 46% in Rustenburg and nearly 50% in Tsantsabane. The corresponding percentages in formal housing are 31%, 32% and 38%. This confirms the importance of informal housing for mobility and, as I have argued, for avoiding the worse consequences of mine closure. It is noticeable that mobility is also prominent in the households in formal houses (see Chapter 5).

We can draw two main conclusions from the data in Table 7.2: informal housing results from structural reasons (poverty) and a need for flexible housing because of the uncertainty associated with mining – realities which policy tends to overlook.

7.4.3 *Informal housing and wealth*

To build on the description above of the social stratification caused by formal and informal housing, we now look at differences in income and expenditure, household assets and self-ranking of wealth.

7.4.4 *Income and expenditure*

Here, we compare informal and formal housing respondents' income and spending and their perceptions of how their income compared to that of other households(see Table 7.3).

Table 7.3 Income, expenditure and income rating for households in case study areas

Income rating and income and expenditure	*Emalahleni*		*Rustenburg*		*Tsantsabane*	
	Formal	*Informal*	*Formal*	*Informal*	*Formal*	*Informal*
Average Likert scale rating, 1 to 5	3.0	2.3	2.6	2.4	2.8	2.5
5: much above average income (%)	2.0	0.0	7.7	10.2	2.9	2.2
4: above average income (%)	14.1	4.5	3.3	0.6	10.3	2.2
3: average income (%)	59.1	21.6	30.9	11.4	51.8	33.8
2: below average income (%)	19.5	31.3	25.1	16.9	24.4	26.4
1: much below average income (%)	8.0	42.5	33.1	60.8	10.7	35.5
Average monthly income, 2018 ZAR values	16,538	6,419	10,285	3,128	9,968	4,980
Average monthly income, 2018 ZAR values (n)	35	17	41	21	45	12
Average monthly expenditure, 2018 ZAR values	11,215	3,925	8,163	3,101	7,328	3,974
Average monthly expenditure, 2018 ZAR values (n)	55	21	54	37	43	19
Average expenditure on food as percentage of total expenditure	20.0	29.1	20.0	27.5	29.5	31.2
Average expenditure on food as percentage of total expenditure (n)	33	19	47	33	39	18

Source: Household surveys.

Emalahleni respondents had the highest household income in both formal and informal houses, at around R16,500 and R6,400 per month. The corresponding figures for Rustenburg and Tsantsabane were much lower, at around R10,000 and R3,000, and R10,000 and R5,000. The stark income differences between formal and informal households once again expose the social stratification. The same patterns are evident for household expenditure.

Households in informal housing spent a larger percentage of their incomes on food. This indicates high poverty levels. However, the percentage of about 30% for households in informal housing in our samples is not extraordinarily high (see Martins, 2005).

The average income and expenditure figures hide important patterns that come to light when we look at the respondents' ratings of their own income as compared with what others in their community are earning. The highest rating was from Rustenburg, with 10% of respondents in informal housing rating their income as being much above average. The percentage is higher than in any of the three case study areas for either formal or informal housing. This suggests that not all informal housing results from structural problems like poverty but can indicate a need for flexibility.

7.4.5 Household assets

I used the asset index to show the differences between the number of household assets (out of a list of 15) owned by people in formal and informal housing.

Figure 7.3 shows that households in formal housing have far more household assets than those in informal housing. In Emalahleni, households in formal housing

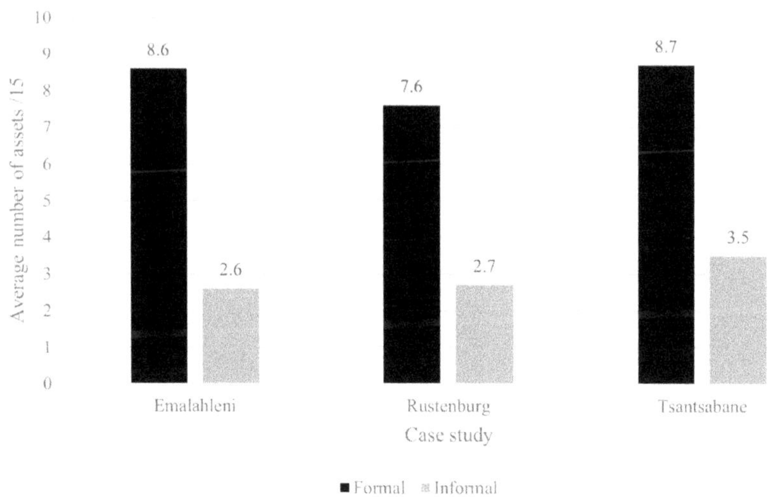

Figure 7.3 Average number of assets for formal and informal houses in the case study areas.

Source: Household surveys.

have four times as many assets as those in informal housing, and the differences are also large in Rustenburg and Tsantsabane. Households in formal houses in Rustenburg have fewer assets than those in formal housing in the other areas. One reason for this may be higher levels of continued migrant labour in this area (see Chapter 5), and many of these households have a second home where they also have assets. Of the informal housing respondents, Tsantsabane's had the highest average number of assets, perhaps because of the high percentage of mine-employed households living in informal settlements in this area. Good mine salaries give these households the means to purchase more household assets.

Once again, these findings illustrate social stratification and point to structural reasons like poverty and poor infrastructure for informal housing and other factors such as migrant labour.

7.4.6 *Self-ranking of wealth*

Here we look again at the wealth self-ranking (Chapter 5, Figure 5.3) to see the differences between households in formal and informal housing and how they have changed over time.

Figure 7.4 shows that respondents in formal housing rank their wealth higher than do respondents in informal housing. The largest difference is in Emalahleni. The reason why Emalahleni respondents in formal housing recorded the highest wealth ranking is that mineworkers' salaries were the highest there. The relatively small difference in Tsantsabane reflects the larger percentage of mineworkers in informal housing. Respondents in informal houses in Tsantsabane rank their

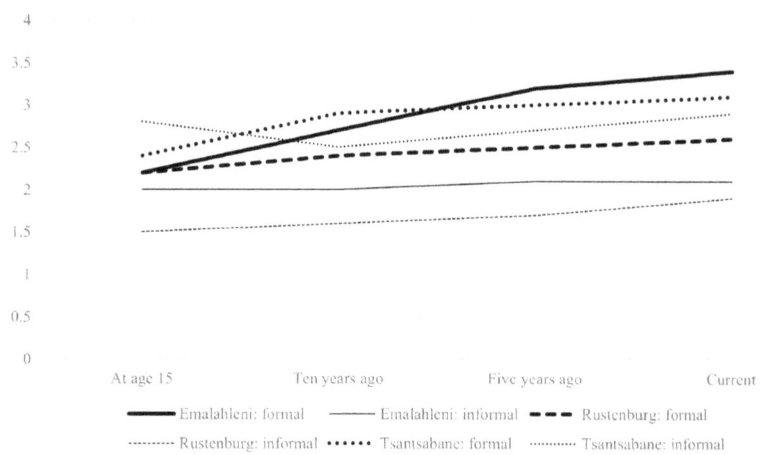

Figure 7.4 Self-ranking of wealth by respondents in formal and informal houses in the case study areas.

Note: 'Current' means at the time of the survey: 2015 for Tsantsabane, 2017 for Emalahleni and 2018 for Rustenburg.

wealth higher than do respondents in formal houses in Rustenburg. The reason for this may be the large scale of instability in Rustenburg over the past decade.

7.4.7 *Place attachment*

Chapter 5 provided a detailed analysis of factors contributing to place attachment. Here, I compare the place attachment of respondents in formal and informal housing (see Table 7.4).

We would expect people living in informal housing to be less keen to stay in their current town than people in formal housing, and we do indeed see this in Emalahleni and Rustenburg. In Emalahleni, 71% in formal houses had a strong or moderate preference to stay as opposed to only 22% in informal houses, and the corresponding percentages for Rustenburg were 66% and 26%. The pattern is very different in Tsantsabane, where 88% in informal houses had a strong or moderate preference to stay, indicating high levels of place attachment, even higher than the 78% for formal houses. Almost half the respondents in informal houses in Emalahleni and Rustenburg were very keen to leave: 46% in Emalahleni and 44% in Rustenburg. Only 2% in Tsantsabane were very keen to leave.

We can draw two conclusions from this. The results from Emalahleni and Rustenburg show the link between informality and mobility. Tsantsabane is probably different because of its lower levels of migrant labour from outside the province.

7.4.8 *Informal housing and social disruption*

Chapter 5 discussed factors influencing social disruption. Here, we look again at the results of the lost wallet question and the crime index (see Chapters 5 and 6), this time distinguishing between informal and formal housing. Table 7.5 shows the levels of belief in the likelihood of someone returning a lost wallet containing ZAR 200, as expressed by these two housing groups.

Table 7.4 Place attachment of respondents in formal and informal houses

Income ranking	*Emalahleni*		*Rustenburg*		*Tsantsabane*	
	Formal	*Informal*	*Formal*	*Informal*	*Formal*	*Informal*
Strong preference to stay	38.5	7.4	47.7	18.2	57.0	78.3
Moderate preference to stay	32.2	15.6	18.6	7.6	20.6	9.9
Unsure	10.2	17.0	9.9	14.7	11.5	7.5
Moderate preference to leave	11.3	14.1	11.0	15.3	5.2	2.4
Strong preference to leave	7.9	45.9	12.9	44.1	5.7	2.0

Source: Household surveys.

Table 7.5 Likelihood that a wallet would be returned

Not likely at all that it would be returned by…	*Emalahleni*		*Rustenburg*		*Tsantsabane*	
	SCB	*Stranger*	*SCB*	*Stranger*	*SCB*	*Stranger*
Formal housing	84.1	91.7	84.5	94.2	84.8	91.7
Informal housing	60.1	92.3	94.1	97.7	59.9	92.3

Source: Household surveys.
Note: SCB = someone close by.

Table 7.6 Average crime index scores

Housing attributes	*Emalahleni*	*Rustenburg*	*Tsantsabane*
Formal housing	19.9	20.1	19.6
Informal housing	20.7	23.7	21.4

Source: Household surveys.
Note: To assess people's experience of crime in the three study areas, the survey asked respondents to rate their perceptions of the levels of six types of crime in their area: burglaries, muggings or thefts; violence between members of the same household; violence between members of different households; gangsterism; murder, shootings or stabbings; and drug or alcohol abuse. A crime index was developed by rating the responses on a scale of 1 to 5, with 4 being 'fairly common' and 5 'very common'. The average was then calculated by summing the responses for each of the six types of crimes.

In Emalahleni and Tsantsabane, more people in formal than in informal houses thought the wallet would be not returned by someone close by: 84% in formal houses in Emalahleni, as opposed to 60% in informal houses, and 85% in formal houses in Tsantsabane, as opposed to 60% in informal houses. In Rustenburg, the position was reversed: 85% in formal housing thought the wallet would be not returned by someone close by, as opposed to 94% in informal houses. Yet when asked about the likelihood of the wallet being returned by a stranger, in all three case study areas more people in informal houses thought it unlikely. These results point to the instability in Rustenburg and the social friction in some of these informal settlements. People in informal housing appear to have less trust in people from outside.

Table 7.6 shows the scores for the crime index for respondents in formal and informal houses in the three case study areas.

Overall, informal areas scored higher than formal areas on the crime index. In line with the responses about the lost wallet, respondents in informal settlements in Rustenburg had the highest scores on the crime index. The Rustenburg results show the largest difference between formal and informal housing respondents. Formal housing in Tsantsabane has the lowest score on the crime index. This may again be a result of the mine-provided housing but could also be related to the fact that Tsantsabane is much smaller than the other two case study areas.

7.5 Conclusion

This chapter explored the dual system of formal and informal housing that mining creates. Although this dual system can be seen in most African towns, it is more pronounced in mining towns. The African literature points to mineworkers' need to be flexible and mobile in mining towns as mine closure threatens livelihoods. In South Africa, the scale of informality is smaller in towns experiencing mine decline than in towns where mining is growing, but it is still higher than the average for South African towns.

This chapter makes three main points about informal housing. First, there is evidence that social stratification develops with informal housing. Households in informal housing have lower incomes, overall (but spend a larger percentage of their income on food), rate their income and wealth lower than other people's and own fewer assets than households in formal houses. This social stratification results from poverty but also from government policies that have given mineworkers a choice of types of housing, mining companies withdrawing from housing provision for their workforce (although not in Tsantsabane) and local government's inability to manage the influx of large numbers of people.

Second, although the inequality is a concern, it is important to understand that informality has a plus side: it gives residents of a mining town, particularly mineworkers, a platform from which to cope with the inherent uncertainty and temporariness of the mining industry. It is difficult to make universal claims about informal settlements in mining towns because the issue is complicated by migration and mobility. The three case studies on which this chapter is based show that informality goes hand in hand with being flexible and mobile. For example, respondents in informal housing in Emalahleni and Rustenburg expressed a strong preference to leave those towns. But I found the opposite in Tsantsabane, where respondents in informal housing expressed strong attachment to their town.

And third, households in the three towns had different experiences of social disruption and crime. The evidence showed that levels of disorder and crime were higher in the informal housing areas in Rustenburg. The picture was of a group of people who mistrust outsiders. The high level of social disruptions and mistrust of outsiders in Rustenburg could perhaps be traced back to the Marikana incident of 2012 or it could have developed since then.

An important implication of the evidence provided in this chapter is that informal housing plays a crucial role for many households in providing flexibility and mobility in the case of mine closure. The temporary nature of mining requires flexible responses, which informal housing does help to deliver. One of South Africa's policy failures has been the inability to apply the informal settlement upgrading programme in mining areas. The income restrictions on beneficiaries are probably the main reason and require reconsideration for mining settlements.

Informal housing often represents an attempt by the householder to deal with the temporary nature of mine work and to avoid becoming fond of the place. This reality stands in contrast to government policies that promote stability, homeownership and place attachment.

References

Amofah, S., Agyare, L. & Bioh, E., 2021. Paradox of informal luxury housing boom in a post-mining town in Ghana. *Journal of Resources Development and Management*, 73, pp. 29–38.

Bezuidenhout, A.B.S., 2015. From compounded to fragmented labour: mineworkers and the demise of compounds in South Africa. *Antipode*, 43(2), pp. 237–263.

Bezuidenhout, A. & Buhlungu, S., 2015. Enclave Rustenburg: platinum mining and the post-apartheid social order. *Review of African Political Economy*, 42(146), pp. 626–544.

Bryceson, D., Jønsson, J. & Shand, M., 2020. Mining mobility and settlement during an East African gold boom: seeking fortune and accommodating fate. *Mobilities*, 15(3), pp. 446–463.

Crush, J. & James, W., 1991. Depopulating the compounds: migrant labor and mine housing in South Africa. *World Development*, 19(4), pp. 301–316.

Gough, K. & Yankson, P., 2012. Exploring the connections: mining and urbanisation in Ghana. *Journal of Contemporary African Studies*, 30(4), pp. 651–668.

Gough, K., Yankson, P. & Esson, J., 2019. Migration, housing and attachment in urban gold mining settlements. *Urban Studies*, 56(13), pp. 2670–2587.

Grätz, T., 2013. *The 'Frontier' Revisited: Gold Mining Camps and Mining Communities in West Africa*. Berlin: ZMO (Zentrum Moderner Orient) Working Paper 10.

Hamann, R., 2004. Corporate social responsibility, partnerships, and institutional change: the case of mining companies in South Africa. *Natural Resources Forum*, 28(4), pp. 278–290.

Hamann, R., 2009. Who takes responsibility for the informal settlements? Mining companies in South Africa and the challenge of local collaboration. In: E. Raufflet & A. Mills, eds. *The Dark Side: Critical Cases on the Downside of Business*. London: Routledge, pp. 125–147.

Hamann, R. & Kapelus, P., 2004. Corporate social responsibility in mining in southern Africa: fair accountability or just greenwash? *Development*, 47(3), pp. 85–92.

Hilson, G., 2013. 'Creating' rural informality: the case of artisanal gold mining in sub-Saharan Africa. *The SAIS Review of International Affairs*, 33(1), pp. 51–64.

Jønsson, J.B.D., Kinaboc, K. & Shand, M., 2019. Getting grounded? Miners' migration, housing and urban settlement in Tanzania, 1980–2012. *The Extractive Industries and Society*, 6(3), pp. 948–959.

Jønsson, J. & Bryceson, D., 2017. Beyond the artisanal mining site: migration, housing capital accumulation and indirect urbanisation in East Africa. *Journal of Eastern African Studies*, 11(1), pp. 3–23.

Lindell, I. & Utas, M., 2012. Networked city life in Africa: introduction. *Urban Forum*, 23, pp. 409–414.

Littlewood, D., 2014. 'Cursed' communities? Corporate social responsibility (CSR), company towns and the mining industry in Namibia. *Journal of Business Ethics*, 120(1), pp. 39–63.

Marais, L., Cloete, J. & Denoon-Stevens, S., 2018. Informal settlements and mine development: reflections from South Africa's periphery. *The Journal of the Southern African Institute of Mining and Metallurgy*, 118, pp. 1103–1111.

Marais, L., Denoon-Stevens, S. & Cloete, J., 2020. Mining and urban sprawl in South Africa. *Land Use Policy*, 98, p. 103953.

Marais, L., Ntema, J., Cambell, M., Cloete, J. & Lenka, M., 2022. Informal settlements in the mining context. In: L. Marais, P. Burger, M. Campbell, S. Denoon-Stevens & D.

van Rooyen, eds. *Coal and Energy in South Africa. Considering a Just Transition*. Edinburgh: Edinburgh University Press, pp. 73–87.

Marais, L., Ntema, J., Cloete, J. & Lenka, M., 2021. Mine housing, assets and informality in Rustenburg: Implications of mine closure. In: L. Marais, M. Campbell, S. Denoon-Stevens & D. Van Rooyen, eds. *Mining and Community in the South African Platinum Belt: A Decade after Marikana*. New York: Nova Science Publishers, pp. 141–158.

Marais, L. & Venter, A., 2006. Hating the compound, but ... Mineworker housing needs in post-apartheid South Africa. *Africa Insight*, 36(1), pp. 53–62.

Martins, J.H., 2005. The household food budget of the wealthy and the poor in South Africa. *Journal of Family Ecology and Consumer Sciences*, 33, pp. 37–45.

Mujere, J., 2020. Unemployment, service delivery and practices. *Critical African Studies*, 12(1), pp. 65–78.

Mwango, M., 2006. *Informal Housing Delivery: Assessing Its Potential Contribution in Formulating Enabling Low-Income Housing Standards: The Case of Kamatipa Settlement in Kitwe, Zambia*. Durban: University of KwaZulu-Natal.

Naumann, C. & Greiner, 2017. The translocal villagers: mining, mobility and stratification in post-apartheid South Africa. *Mobilities*, 12(6), pp. 875–889.

Nkomo, M., 2014. 'Beyond the compound': organisation of social life in an informal settlement within a South African mining town. *Labour, Capital and Society*, 48(1–2), pp. 192–211.

Rubbers, B., 2021. Mining towns, enclaves and spaces: a genealogy of worker camps in the Congolese Copperbelt. *Geoforum*, 98, pp. 88–96.

Rubin, M. & Harrison, P., 2016. An uneasy symbiosis: mining and informal settlement in South Africa with particular reference to the Platinum Belt in North West Province. In: L. Cirolia, T. Görgens, M. van Donk, W. Smit & S. Drimie, eds. *Upgrading Informal Settlements in South Africa. A Partnership-based Approach*. Cape Town: UCT Press, pp. 145–174.

Stark, L. & Teppo, A., 2021. Examining power and inequality through informality in urban Africa. In: L. Stark & A. Teppo, eds. *Power and Informality in Urban Africa: Ethnographic Perspectives*. London: Zed Books, pp. 1–19.

Stats SA, 1996. *Census Data, 1996*. Pretoria: Stats SA.

Stats SA, 2001. *Census Data, 2001*. Pretoria: Stats SA.

Stats SA, 2011. *Census Data, 2011*. Pretoria: Stats SA.

Stats SA, 2016. *Community Survey, 2016*. Pretoria: Stats SA.

Stewart, T. & Drewes, E., 2018. The Khumani approach to homeownership in Postmasburg. In: L. Marais, P. Burger & D. Van Rooyen, eds. *Mining and Community in South Africa: From Small Town to Iron Town*. London: Routledge, pp. 157–170.

Steyn, G. & Atamelang, K., 2014. Letlhakane, Botswana: informality in transition. *International Journal of Housing Science*, 38(1), pp. 43–51.

Van Bockstael, S., 2014. The persistence of informality: perspectives on the future of artisanal mining in Liberia. *Futures*, 62, pp. 10–20.

Van der Watt, P. & Marais, L., 2019. Normalising mining company towns in Emalahleni, South Africa. *The Extractive Industries and Society*, 6, pp. 1205–1214.

8 Findings, contributions and policy recommendations

8.1 Introduction

I started the book by referring to path dependence, using the analogy of the QWERTY keyboard. Millions of people today still use the QWERTY arrangement created in 1873, even though a different arrangement of the letters might make it possible to type more efficiently. The cost and inefficiencies of changing the keyboard outweigh the possible benefits.

We see a somewhat similar situation in towns that have become dependent on mining. The literature has been particularly concerned with the social disruption brought about by mining booms. In this book I acknowledge this concern, but I argue that perhaps a more important effect of a mining boom is the dependence it creates. This dependence develops after the initial social disruption. The disruption wanes, but the dependence remains. People expect the prosperity to continue. On the whole, the government and the mining companies do not think about decline, although mining companies have started to include closure in their plans during the past two decades. Once mining decline sets in, the dependence makes it difficult for people and organisations to change. Often, mining keeps this dependence alive through periods of boom and bust. Mining also creates material dependencies, in the form, for example, of environmental damage that makes it difficult to plan for alternative economic activities, or mine infrastructure that is difficult to adapt for other uses.

In this book, I have looked at global and South African practices to see how mining creates dependence and why governments, mining companies and households should try to minimise the dependence so as to minimise the risks attendant on closure. I have pointed out how housing policy during a mining boom can create long-term dependence. South Africa's housing policy emphasises homeownership, wealth creation and local development. Homeownership is intended to help households accumulate assets and to ensure stability through place attachment.

The place attachment literature generally views the concept of place attachment positively. It is seen as having economic and psychological benefits. New large-scale mining and energy developments often disrupt the place attachment of smaller and remote areas. Social disruption is, therefore, a central concept in

DOI: 10.4324/9781003262015-8

the mining literature. Yet the literature seldom looks at the place attachment that mining creates, or questions it. In South Africa, housing policies for mining areas emphasise the importance of housing for creating place attachment, which they see as central to creating stability. Housing policy for mining areas in South Africa promotes stability through place attachment and emphasises the role of mining in development. The policy assumes a viable long-term economy and ignores the reality of mine closure.

I analysed the concepts of place attachment and asset-building in Chapter 2. Both concepts are important for creating stability and fostering long-term thinking. Empirical research shows that homeownership and length of stay support the development of place attachment. Yet, little research has been done on the possible adverse outcomes of place attachment. As mining is a temporary activity, the emphasis on permanence may be inappropriate in mining areas. In Chapter 2, I also discussed the concept of social disruption. Although researchers have used the concept widely to explain the adverse social implications of mining developments, they have not applied it to mine closure, but focused on mining as a disrupter during mining booms. This is partly because most of this literature originates from the Global North and finds little indication of social disruption at mine closure (O'Connor & Ruddell, 2021). In this book, I view mining as an economic activity that creates a dangerous dependence. In my view, mine closure becomes the disrupter.

Responses to mine closure and the social consequences of closure were the topic of Chapter 3. Despite a growing body of research on the social consequences of mine closure, the literature seldom links mine closure to social disruption. The front-end approach of the mining industry results in large investments to obtain a social licence to mine. In this chapter, I also discuss the concept of shrinking cities and planning for decline. The literature on shrinking cities emphasises accepting decline and not necessarily opting for economic diversification. Economic diversification programmes are not the only possible response to decline. Instead, cities experiencing economic decline should learn to value it. The benefits of shrinking include a reduction in energy and water use, fewer traffic problems and the opportunity to reconnect the city with the natural environment. Many mining cities and towns can learn from the shrinking cities literature. Not only should they value decline, they should plan for decline from the outset. Examining the possible benefits of decline could replace overemphasising economic diversification.

In Chapter 4, I analysed the policies and programmes that apply to mining towns in South Africa. Most of these emphasise the importance of mining for development. Although this link is important, overemphasising or only emphasising this link is risky. South African policies aim to phase out migrant labour and instead promote homeownership and housing programmes for stabilising mining towns. However, these policies could create new dependencies. They ignore existing dependencies, such as the persistence of migrant labour. Mine closure could disrupt the stability created by housing and homeownership programmes.

Chapter 5 looked at the continuation of migrant labour and the mobility associated with mining as a path dependence. Housing programmes that emphasise homeownership and stability ignore continued migrant labour and mobility. The chapter also analysed the international and African literature linking migration patterns, mining development and mine closure. I used the example of copper mining in Zambia to challenge the simplistic linking of mine development, housing, homeownership and modernity. I argued that continued migration and mobility could contribute to development and act as a mitigating factor for households at mine closure. The empirical evidence from the case studies of Emalahleni, Rustenburg and Tsantsabane points to five main findings. First, compared to historical migrant labour patterns, the level of migration is much lower today. This is because of local labour recruitment, the dismantling of the compounds, the promotion of homeownership and the focus on housing choice in government policy. The policy has helped to create place attachment, which reduces migration. Second, despite this policy, migrant labour undeniably continues, most prevalently among the older and less educated mineworkers (Marais et al., 2021c). Consequently, as mines mechanise and the older mineworkers retire, the levels of migrant labour will decrease. However, I believe it will not die out completely. Third, large numbers of people migrate to the mining areas in search of a job or to benefit from the secondary economies around mining. Many of these migrants live in informal or rental housing. Fourth, the evidence shows that migration can contribute to asset-building. When asked to self-rank their wealth, households who had migrated to the mining areas estimated that it had improved substantially over the past 15 years. Finally, homeownership and high levels of housing satisfaction evidently created place attachment, while low levels of housing satisfaction were significantly associated with higher levels of experiencing crime. The question is who is most at risk when a mine closes. I think it is those who are employed on the mine, own a house and are paying off a mortgage. Those who rent may be comparatively unscathed. They will be flexible enough to relocate without risking their asset base.

Chapter 6 considered the implications of mine closure for mining towns in Africa. The African Mining Vision emphasises the importance of mining for creating development and modernity. In contrast, the African literature points to the devastating local consequences when mines close. Many of these mining towns struggle to cope. Many of the consequences of mine closure result from the dependence mining creates by providing housing, infrastructure and welfare. Mine closure leads to infrastructure decay, maintenance problems and rapid economic decline. Considering the likelihood of decline, the second part of the chapter analysed housing and wealth creation in the three case study areas. There are four main types of housing in the sample: homeownership for long-standing residents, homeownership for mineworkers, rental housing for mineworkers and rental housing for newcomers. I showed how mining does develop housing, wealth and place attachment. But I also showed how some mineworkers prefer not to risk accumulating housing assets and developing place attachment. The most at risk are mining households who have bought their own houses and still have mortgages to pay off – especially if they bought houses in the past five years. Those least at risk are the mineworkers renting. They can give notice and then be mobile and migrate. Although mine closure holds risks for landlords, the scale of the risk is smaller than for those caught up in a system that functions entirely on ownership and mortgages. Kumba's provision of rental housing in Tsantsabane is an example of a mining company playing an active role. The empirical evidence shows that this is very good family housing and does enable households to accumulate assets. Rental housing can help create stability.

In Chapter 7, I looked at the place of informal housing in the mining context. Mining and informality go hand in hand in Africa. In South Africa, mining towns have higher levels of housing informality than other towns. Many have a dual formal and informal housing system. Mine closure often reinforces this pattern, with informality remaining high even after closure. Governments generally view informality and continued migration as contrary to the ideals of stability and modernity. Although such an argument is not without merit, it is not the only valid argument, for two reasons. First, informal housing springs up because of a policy that emphasises choice and because mining companies do not provide adequate rental housing. I argue that informal housing is a consequence of the current overemphasis on homeownership and assets in mining areas. Second, informality in mining towns helps households deal with the temporariness of mining. Migration is a way of dealing with uncertainty and possible closure. The analysis of informality in the three case study areas supports my argument that not all informality is inappropriate.

8.2 Theoretical and conceptual contributions

This book makes several theoretical and conceptual contributions: on mining dependencies, on place attachment in mining towns, on assets and stability in mining towns and on economic diversification and planning for decline. There has been very little theoretical or conceptual thinking outside theories that view mine closure as an economic transition since Digby (2012) linked mine closure and the economic transition literature. The problem with economic transition theories is that they focus largely on the consequences of mine closure and finding alternative economies. This book remedies this deficiency. In doing so, it makes a substantial contribution to the mine closure literature.

8.2.1 Mining dependencies

My central argument in the book is that mining booms create dependencies. This argument stands in contrast to two commonly held views in the literature: mining disrupts (Brown et al., 2005; Ruddell et al., 2014) and mine closure does not disrupt (O'Connor & Ruddell, 2021). The remote locations of many mining areas exacerbate these dependencies. Because mining regions have few other economic activities, remote locations welcome mining investments and develop local strategies dependent on the mining economy. Many households do the same. They buy houses and other assets without considering mine closure. The mining literature emphasises the social disruption that new mining investments bring. However, the social disruption abates over time. What starts as disruption becomes a dependency. Yet, the literature seldom considers the new status quo as a dependency.

I emphasise two main dependencies associated with mining: path-dependent policies and material dependencies. When a mine opens, the main question for a mining company is, what should they do to keep their mining licence? A substantial body of work has investigated the social licence to operate (Owen & Kemp, 2013; Koivurova et al., 2015; Matebesi & Marais, 2018; Matebesi, 2020). Despite the value of this licence, excessive focus on it can lay the foundation for long-term dependencies between mines and local communities. Another dependence comes from housing policy overemphasising homeownership and stability and ignoring the value of the existing migration and mobility. Homeownership and stability could inhibit mobility and lock-in some households, making outmigration at closure difficult. The policy dependencies also arise from local strategies based on continued economic growth and mining prosperity. Generally, local plans show little understanding of how to plan for closure. Closure is largely an unthinkable idea in the local plans of mining regions. The uncertainty associated with mining requires that local strategies consider more than one scenario. Nel and Oranje (2022) refer to this as 'planning in the dark'. Moreover, it requires a strategy linking growth and decline. For example, how can a strategy conceived during a boom mitigate the consequences of decline or how can a boom be planned so that closure will be less disruptive? These questions are seldom asked in the mining context. As I have noted in Chapter 3, a front-end approach dominates. This approach largely focuses on creating stability, obtaining local consent and minimising disruption so the mine can get going. Local strategic planning should focus more on mitigating the consequences of decline by not creating dependencies.

Material dependencies affect local communities for a much longer period than the mining. These dependencies are often environmental. Rehabilitation of mining land can be difficult. Acid mine water can make land unsuitable for farming. The dependencies result from cumulative impacts while mining takes place and can make it hard to develop alternative economic functions for decades and even centuries (Esterhuyse & Buscke, 2022).

Material and policy dependencies can be intertwined. A good example is the Free State Goldfields, where garden city planning principles applied in the 1950s are now making it hard to manage decline (Marais & Nel, 2016). The wide roads and extensive infrastructure are examples of material and policy dependencies. Maintaining this extensive infrastructure after mine decline is a problem. Local strategies that create material dependencies place an extra burden on local decision-makers in periods of decline. The question remains, how to develop appropriate infrastructure that serves the immediate needs of mine growth and yet can be either shut down or reused after mine closure?

In Chapter 4, I highlighted another type of dependency: goal dependency. In its attempt to redress the harm done by the inhumane compound living system, South African policy pays little attention to the benefits of rental housing for individuals or accommodation provided by employers. Redress implies getting rid of the compounds and migrant labour. This has made the policy goal-dependent and obscured the reality of continued migrant labour and mobility. Redress, like the social licence to operate, only looks at the immediate need and does not consider a future in which mine closure is likely. The evidence from Chapter 5 shows that migrancy can, in fact, create assets and allow mobile households flexibility at closure.

There is more to this goal dependency. As noted in Chapter 4, mining companies, governments and unions all agreed to get rid of long-term mine housing liabilities. They did not want to be responsible for any housing when a mine closes. The government wanted to offer housing choice. The unions wanted to give their workers a choice but saw the financial benefit of workers obtaining a living-out allowance. The result has been a housing system entrenched in neoliberal thinking supported by all parties. This approach has become embedded in the South African mining context. The irony is that only the mining companies have considered the long-term implications of mine closure.

8.2.2 *Place attachment, mining towns and dependence*

The development of place attachment in mining towns is an example of dependence. Mining research seldom considers the social disruption that can be caused if people develop place attachment and then find the place they are attached to is in decline. It rarely describes or measures levels of place attachment. Social disruption research assumes existing levels of place attachment. Decisions on whether to set up a company town, establish an open town or promote fly-in-fly-out arrangements result from different assumptions about place attachment, its value or concerns about developing it. Company towns and fly-in-fly-out arrangements prevent the development of place attachment. Open towns and normalisation assume that place attachment will develop and proponents of this approach view place attachment as a good thing.

Against this background, this book makes the following conceptual contributions to the topic of place attachment in mining towns.

First, it highlights the concept of place attachment in mining towns. There is value in highlighting this and giving it a more central role in mining studies. Research on social disruption in mining towns seldom acknowledges that the disturbance can result from place attachment being disrupted. Second, I use the concept of dependency to show that mine housing policies create place attachment. This link between social disruption, place attachment and dependencies is new in the literature on mining communities. Third, in contrast to the international literature that views place attachment only as a positive phenomenon, I argue that on the contrary it can be a problem. It is a problem in places where the economic foundation is unstable or does not provide for long-term prosperity. Place attachment in a mining town could be detrimental as it might inhibit mobility at closure. I argue that *lack* of place attachment could be a good thing. For example, the migration and mobility associated with the lack of place attachment might be a way for households to mitigate the effects of closure. I also question the simplistic link between homeownership, asset creation, stability, place attachment and development. I do not dispute that such a link exists, but I question whether this is the only link that policies should consider. The empirical evidence from the three case studies shows that continued migration and rental housing can encourage stability, asset creation and development.

8.2.3 *Housing assets, stability and closure*

Policies promoting mining towns assume that permanency, assets and homeownership create stability. The stabilisation of mining towns through housing and homeownership underlies several South African policies, as explained in Chapter 4. The policies focus on open towns and homeownership at the expense of rental housing. Researchers often blame the 2012 Marikana tragedy on poor living conditions and low levels of stability (Marais et al., 2021a). However, I believe the arguments are more complex and less linear. For example, fly-in-fly-out arrangements show that mining households do not need to situate their housing assets near the mines. Despite some adverse social and economic implications of the fly-in-fly-out system (Haslam McKenzie, 2020), it does show that mobility has value and need not impede the development of assets. The evidence in Chapter 5 shows that mobility can create stability and modernity. Stability is not inevitably linked to homeownership and housing assets near the mines. My argument is that encouraging settled dwelling near the mines could develop long-term dependencies. The danger is that stability of that kind can become rigidity when a mine closes. Mobility should be an option while the mine is operating and a possible strategy for closure.

Housing assets can create problems not only for their owners but also for local authorities when mine decline sets in. Municipalities become dependent on the rates and taxes from the properties under their jurisdiction. An outflow of people will reduce this income. Mine closure might disrupt what these local authorities originally saw as a stable asset base.

8.2.4 *Planning for decline and the elusive nature of economic diversification*

Policymakers' usual strategy for avoiding dependence on mining is economic diversification. They count on this to ensure long-term economic prosperity for a town. Logically, once diversification shifts the local economy away from mining, there is no reason why they should not promote homeownership, place attachment and stability. But their argument hinges on economic diversification being feasible.

In mining towns it has been difficult. Success stories are few. In practice, economic diversification remains an elusive concept for mining towns worldwide. There are several reasons for this. Many are in peripheral economic areas, with limited national and international economic links or agglomeration economies outside their mining economies. Furthermore, although mining usually improves local infrastructure, the infrastructure is directly related to mining and is difficult to adapt for other industries. For example, the railway between Kathu and Saldana Bay was designed mainly to carry iron ore. This material dependency inhibits economic diversification. Another problem is that mining seldom offers the foundation for developing related industries. A body of research on economic diversification in remote areas says the diversification usually happens in sectors related to the original industry (Boschma, 2017; Fitjar & Timmermans, 2019; Wixe & Andersson, 2017). For example, mobile phone technology improved in Finland because of the forestry industry that required an appropriate communication method. The problem with the mining industry is that related industries are hard to find and mining skills are often unrelated to tertiary economic activities. Mining often benefits 'gateway' cities more than the cities or towns where mining occurs (Atienza et al., 2021). Consequently, the mining localities seldom benefit from the upstream and downstream economic activities of mining, such as financing and legal services. These activities usually occur in one or two gateway cities in a country – usually where the head offices of the mining cities are situated.

These four obstacles – remote locations, mining-specific infrastructure, unrelated industries and the dominance of gateway cities – make economic diversification highly unlikely in many remote areas. Yet when mining towns in these areas plan, they usually envisage economic diversification. The planners often use the example of Johannesburg, which started as a mining town and later developed a diverse economy. Successful economic transformations away from mining are rare. One example in South Africa, other than Johannesburg, is Newcastle. This town managed to diversify its economy largely because of government subsidies and a proactive chamber of commerce (Marais et al., 2021b).

The question is what the chances of economic diversification are for my case study areas. Rustenburg and Emalahleni have a better chance than Tsantsabane. Both are located on the periphery of Gauteng, South Africa's economic heartland. Better economic connections between these two cities and Gauteng could buffer mine closure. Nevertheless, mine closure and decline will cause them substantial economic shocks. Both places should foster the links with Gauteng and think about planning systems that reduce place attachment and plan their cities to ease the disruption of closure. Tsantsabane (formerly Postmasburg) is much smaller and more remote. Despite several initiatives from the mining companies to support local value chains, the mining economy remains dominant and economic diversification highly unlikely. Linking the mining value chain with local people might increase their dependence on mining.

If economic diversification is a forlorn hope and it would make better sense to accept decline, the question then is what kind of housing policy to adopt for mineworkers. In Tsantsabane, Kumba is dealing with the risk by providing only rental housing, not ownership. However, Assmang has opted for ownership in Tsantsabane. I believe Kumba's is the appropriate response, despite pressure for ownership. If the mine closes, employees in rental housing will find it easier to relocate. It also means that Kumba has taken responsibility for the long-term liability of managing these houses in Tsantsabane. Although nobody knows what the consequences of closure will be, this seems a more rational approach to handling the liabilities of mine decline.

Beyond the direct implications for housing policy, cities and towns that experience mine closure should also consider the lessons from shrinking cities. Managing shrinking and valuing it is a more sensible way of dealing with decline than working on the assumption that growth will continue. Unfortunately, this approach does not yet seem to have caught the attention of South Africa's policymakers. The lessons from the shrinking cities literature go beyond planning. The innovative step is to value decline and see what benefits it can bring rather than treat it as a disaster. The question is whether one can plan mining cities to make decline easier.

8.2.5 *Expanding the literature on mine closure*

The book contributes to the mine closure literature. Although work is increasing on the social aspects of mine closure (see Chapter 3), the overall body of work remains small. I have several concerns about this research, but here I will mention two main concerns: there is insufficient theoretical work and most of the work is descriptive.

The mine closure literature has been linked to the literature on economic transitions (Bainton & Holcombe, 2018). The shift away from coal is an example of such a transition (Swilling et al., 2016; Marais et al., 2022). Emalahleni is an example where the economic transition literature is relevant. With the transition away from coal being urged on countries because of global climate agreements, the workers' needs must be attended to. However, not all mine closures are related to economic transitions; they can be the result of diminishing resources or changing market conditions. There are examples of economic transitions outside mining. Economic transition in the Rust Belt states in the US (away from manufacturing) is an example. Many economic transitions in Europe involve large subsidies by the government over a prolonged period. Such examples are often used to inform economic diversification strategies, ignoring the differences of context and complexity. In this book, I deliberately avoid suggesting economic diversification as a solution for mine decline.

I contribute to the mine closure literature by assessing existing policies and practices that will make mine closure difficult. The emphasis is on understanding the current approaches and the dependencies they cause. The book expands the descriptive literature on mining booms, mine closure and mining policies, but it complements the description with analysis of the implications of closure. Although mining companies have improved their thinking about mine closure, this thinking has not yet filtered into communities and their organisations' strategic plans. To some degree, the opposite has happened. Current practices create more dependencies, and rather than questioning them, people become still more dependent on them. My approach in this book has been to challenge this approach and foreground the closure problems that mining processes create.

This book's most important contribution to this literature is its innovative approach. Rather than describing the social consequences of mining in towns where mining is in decline or where mines have closed, I have largely investigated towns where mining is still a major economic sector and still growing. I looked at the profiles of these places and their people to identify future risk factors. The challenge in planning and practice is precisely this: to identify the risk factors and dependencies *while mining is growing*, to plan for the likelihood of mine decline.

8.3 Policy recommendations

I do not recommend inflexible one-size-fits-all policies. Every mining town is different. The differences need to be taken into consideration in the following recommendations. In essence, my recommendations challenge simplistic assumptions about the link between mining and development. I want to broaden the debates about economic growth, stability and diversification. I agree that the relationships between these factors should receive attention in the policy, but they are not the only ones to be considered. I am more concerned about policymakers' oversimplified ways of addressing mine decline. We need alternative approaches.

The chapters in this book have criticised the dominant approach of promoting homeownership for mine housing. I do not say that homeownership per se is inappropriate; my concern is that other forms of tenure do not receive equal attention. Why has employer-provided housing largely disappeared from the housing mix? Any policy approach should have a tenure mix and encourage mining companies, municipalities and households to facilitate the development of such a mix. Mineworkers and other inhabitants of mining towns will make their own decisions. But policy, municipalities and mining companies could ensure that several options are available.

8.3.1 Responsibility for housing in mining areas

Housing for mineworkers is an issue worth investigating. Some would argue that the private sector should take full responsibility for mine housing. I see several reasons why mining companies and local governments should take responsibility: to redress historical injustices, to deal with the problems of remote locations, to improve health and well-being and productivity and to mitigate the risks attendant on mine closure.

The history of migrant labour and compounds requires redress. Clearly, the policy must help with redress. Compounds are no longer an option. Individual housing choice is at the heart of the current policy, but this has not necessarily created better housing for mineworkers (Marais & Venter, 2006; Pelders & Nelson, 2018). In place of the compound, we have informal housing. Redress also requires a concerted effort to understand mineworkers' needs and migration patterns. Many housing problems in mining areas result from poor local planning and governance and mining companies withdrawing from providing any form of housing. A policy that deals with history requires proactive collaboration between mining companies and local governments. But an approach that understands the history should also consider the continuation of migrant labour and mobility. Although redress is important, policies must not focus only on redress (see Chapter 4).

The remote location of many mining settlements requires a dedicated housing response. Historically, company towns were the mining industry's response to the remote location of mining operations. Mining companies had to attract their workforce by providing decent living conditions. More recently, open town policies and fly-in-fly-out arrangements have attempted to address this reality. Fly-in-fly-out arrangements accept the importance of existing urban agglomerations and settlement patterns. Block roster shifts and high salaries work well in places where mine work is largely outsourced. Open town policies in remote locations, however, are something I have criticised strongly in this book. Providing family housing and homeownership in such locations creates dependencies that put people at risk. Different tenure forms are needed to ensure flexibility. The responses by the two mining companies (Kumba and Assmang) in Tsantsabane show the importance of mining companies' role in the provision of housing.

Getting housing right may increase productivity. There is a well-established link between housing and health and well-being (WHO, 2012). Although there is no empirical evidence, good housing and living environments will probably increase productivity and reduce absenteeism. Mining companies and local governments should get involved because decent housing for mineworkers could improve health and well-being, boost productivity and help to attract skilled labour.

Most importantly, mining companies and local governments must take responsibility for mineworker housing because almost all mines eventually close. Mine closure means that housing finance and planning should consider the likelihood of decline. Mine closure hampered some of the first homeownership programmes that included black people in the Free State Goldfields (Marais, 2013a). The area was redlined by the banks after large numbers of mineworkers in Thabong (the former black township near Welkom) could not repay their mortgages. There is also the danger of low-income households and homeowners being locked into a town where mining has ceased (Ntema et al., 2017).

8.3.2 Housing choice

Good housing policies provide choice. The choice should be available both in theory and in practice. The policy recommendations I make here do not argue against this principle of choice but expand on it. Below, I explain why choice should be part of most housing policies.

Compounds were the primary housing delivery system under apartheid. In Chapter 4, I described the human rights and health concerns associated with the compound, which until the mid-1980s was virtually the only housing for black mineworkers. My criticism of homeownership and open towns is not about the tenure type per se; it is about the lack of choice. The push in current policy is for homeownership. Although not as narrowly prescriptive as apartheid policy for compound living, it does minimise choice.

The evidence in this book has shown that some mineworkers choose housing types other than homeownership and family housing. Households make their own decisions about tenure and housing type on the basis of their understanding of their household and the economic prospects of an area. A substantial number of these mineworkers choose rental housing or informal housing. This means that many existing households extend their house or build backyard rooms to accommodate the demand for rental housing.

However, providing mineworkers with choice does come with some problems: poor quality housing, lack of local government support, irregular land use and discouragement of mine-provided housing. The quality of informal housing is poor. Evidence from Chapter 6 shows that service levels are substandard in some areas. Very few local governments support private rental markets. This support is crucial in mining towns, as formal housing processes are unlikely to cope with the demand that a boom creates. Such support should promote infrastructure that can manage the increased densities, tax incentives for mineworkers' rental housing and building support and guidance. Informal housing often springs up on traditional land where land use regulations are less stringent. In the case of Rustenburg, this land is also near most mines. The difficulty of providing municipal services on traditional land results in poor services. A major problem is that giving the mineworkers a choice appears to absolve mining companies from any responsibilities for their housing.

I argue, to the contrary, that there is indeed room for mine-provided rental housing. This is not a call for a return to the compound. The evidence from Tsantsabane shows that mine-provided rental housing can create family housing and develop assets. Company-managed rental housing could well be part of the offering to mineworkers. The main advantage of this approach is that long-term liability for the houses remains with the mining company.

Housing systems must provide a variety of options. This is important because people's different experiences, household composition, and life choices mean that a one-size-fits-all housing approach will not work. Post-apartheid housing policy is misguided in pushing homeownership as the main option and underplaying other possible and viable housing options.

8.3.3 *The importance of rental housing*

Mining companies and local governments have not fully comprehended the potential role of rental housing. In Tsantsabane, Kumba provides rental housing for its workers. This is a classic example of the role rental housing could play in ensuring flexibility for mineworkers. Tsantsabane's remote location in particular makes this a more appropriate response than promoting homeownership and is an excellent example of mine-provided housing. Although tax on fringe benefits limits the attractiveness of this for the mining companies, it remains a viable option in many remote mining places (Marais et al., 2018). However, the mine-provided rental housing in Tsantsabane is the exception to the rule.

Other forms of rental housing develop naturally and as part of a larger housing market. One of the most common is backyard dwellings for rent. This market caters for people who cannot afford, or do not want, homeownership. There are of course quality concerns. Chapters 5–7 of this book mentioned problems of poor infrastructure access for backyard dwellers. A further problem is that backyard rental housing could put a strain on the existing municipal infrastructure and services. This type of housing does fill a need, but mining companies and local governments seldom acknowledge it or promote it. One obvious advantage is that backyard rentals encourage density and help to prevent urban sprawl. If local governments are to develop an appropriate strategy for rental housing, they need a detailed understanding of its value as well as its negative implications and an appropriate planning response. In all three case study areas, backyard dwelling had developed naturally. Municipalities and mining companies could promote and manage this form of housing.

The national government has a comprehensive rental housing policy and strategy, but it has not contributed to any housing in mining areas (Cloete et al., 2009) (see Chapter 4). This is surprising, as many mining areas have substantial numbers of people who could use this form of housing. There is also potential for partnerships with the mining sector. The income limit set by the government is probably the main reason why this has little attraction in mining towns. Mineworker incomes are too high to use this policy.

8.3.4 *Mine-provided housing and the company town*

Closing or normalising company towns is a prominent government policy worldwide. In South Africa, the policy has shifted from creating company towns to developing open towns. Mining companies and labour unions have endorsed this shift. However, we should be asking whether there is really no value in the company town approach. I argue that it should not be dismissed out of hand. The following are four reasons for keeping this as an option for mine housing.

First, company towns are appropriate for remote mining areas. The original reason the mining companies set up these towns was to provide a decent living environment that would attract skilled workers to remote locations. This reason remains valid for remote mining areas despite rapid changes in mining work. The company town model should remain an option in remote locations.

Second, the alternatives, normalised or open towns, shift the long-term responsibility to the local government. I think the long-term liabilities should be the responsibility of mining companies, or at least shared by them. The question is who takes responsibility for mine housing when the mine closes. The mining company can bulldoze a company town (with permission from the relevant departments) and rehabilitate the land, as they will be doing with the mining land. Rehabilitated land might create new economic opportunities. The revegetation of land on which settlement has taken place could also prevent environmental damage such as by floods. This point aligns with arguments made in the shrinking cities literature (see Chapter 3). Company towns mean that mining companies accept responsibility for short- and long-term benefits and liabilities. The company-provided housing in Tsantsabane is a variant on the company town approach. In this case, the mines developed the houses as part of the municipality's urban system, but the company remains responsible in the long term. Kumba will pay property tax on the houses and land well into the future. Although there are no guarantees of a smooth process when the mines close, and it is unclear what post-closure conflicts may develop, in the end it is the mining company that bears the liabilities.

Third, company towns usually manage fluctuations of boom and bust better than open towns. The large-scale development of informal settlements in all three case study areas shows that local governments find it difficult to ensure adequate housing for the influx of people. Company towns usually use worker camps to deal with the influx during construction and plan their settlements accordingly.

Finally, company towns could have well-being and productivity benefits (Van der Watt and Marais, 2019). Good living environments are important for mineworkers' mental health. This could reduce the number of mining accidents. A company town or mine-provided houses would offer better living conditions than the informal settlements that proliferate in mining town. And there is evidence of company towns encouraging racial integration, Lime Acres, near Postmasburg, being an example.

8.3.5 *Planning for decline*

The mining industry has included closure planning in mine life-cycle planning over the past two to three decades. This is a major step forward in the industry's response to closure. Today, most mining companies have closure plans in place. These plans are usually adjusted annually as technology changes and knowledge about mine closure develops. However, this planning is mainly related to environmental concerns. Rehabilitating land and limiting water contamination are central themes. There are extensive guidelines available to assist (ICMM, 2019). Less prominent in this environmental approach is what decline would mean for mining communities. Mining companies consider this falls outside their ambit. Considering the huge effort mining companies make to obtain a social licence to operate when the mine opens, the lack of measures to deal with the social issues when the mine closes is striking. A further problem is that mining companies seldom think about how their social investments create long-term dependence. I believe planning for decline requires thinking and planning with the following main ideas in mind: inclusivity, planning for closure from the start and valuing decline.

Planning for closure should be inclusive, like the process for obtaining a social licence to operate. Measham and Zhang (2019) have pointed to exclusivity concerns when applying for a social licence. For example, the process tends to exclude women and other vulnerable groups in communities. Because this exclusion occurs during the start of the mining process, it is likely also to occur at mine closure.

Closure planning should consider making decline easier by planning for it at the start of the mine's life cycle. This seldom happens. The large amounts of money available at the beginning of the mining process often result in planners and governance creating dependencies. What would planning for closure at the start mean in practice? It could mean planning the physical and social infrastructure with repurposing in mind, even if this means dismantling infrastructure and rehabilitating housing areas. It could mean using the concept of modular infrastructure, often applied in the ICT sector. Using this concept in civil infrastructure would be more difficult, but the idea should be prominent in planning mining towns. Modular infrastructure makes it easy to dismantle infrastructure that is no longer useful. It could mean that authorities who plan settlements should give consideration to the rightsizing of infrastructure and properties. These ideas require careful consideration by all role players and should not be the responsibility of mining companies alone.

Planning systems should value decline. Decline could have substantial ecological benefits and make it possible to revegetate urban areas (see Chapter 3). The main problem is that both mines and governance bodies find it difficult to acknowledge the social and economic consequences of mine closure. Often, this is ignored or glossed over in a way that does not recognise the reality and hardship that goes with it. The notion of planning for decline and shrinking is probably the most important policy proposal that mining communities and their organisations should embrace.

8.3.6 *The importance of understanding informality*

Chapter 7 showed that informality (i.e. being outside government regulation, control, support and taxation) is common in South African mining town housing. The dual system of housing formality and informality is also noted in the literature on mining in Africa. Planners and city managers often see urban informality as a problem and aim to minimise or eradicate it. The emphasis on homeownership and sustainable human settlements in South African policy (see Chapter 4) represents an attempt to get rid of informal housing. I want to argue that despite some of the negative consequences of informality, it does have a role and policymakers should not view it only as negative.

It might be the sensible choice for some households. Both mining and non-mining households might choose, contrary to government policy, *not* to develop place attachment and *not* to invest as much as possible in the town where they work. In Chapters 4 and 6, I presented an argument for mine-provided rental housing, using the example of Tsantsabane mine housing. Where the mine does not provide rental housing, it is no surprise that some mineworkers choose informal housing. This may not be ideal for health and productivity, but some people prefer it.

The choice of informal housing may be a way of dealing with uncertainty. The evidence in Chapter 7 shows that many people in informal housing are low-income newcomers to the three case study areas. Fairly small numbers of people in this type of housing are mineworkers. Yet, many of the people in both types of household are migrants to the mining areas. Informality provides them with a foothold but also allows easy outmigration if needed.

Informality originates from the inability of local governments to plan adequate housing. When I argue for a broader understanding of informality, I am not suggesting that local governments should not be responsible for housing provision; I am suggesting that informal settlement upgrading should be part of their housing strategy in mining areas. At the time of writing, very little informal settlement upgrading was being done in mining towns (Huchzermeyer, 2021). The income limit of less than ZAR 3,500 per month to qualify for informal settlement upgrading may be a stumbling block. The appropriate application of the informal settlement upgrading programme would create a policy approach where place attachment and the building of assets are not the main aim.

Finally, in mining towns like Rustenburg the relationship between tribal land, informality and service provision remains a problem. Formal planning on tribal land is difficult, not just because of technical difficulties but also because of power play between local authorities and traditional authorities. South Africa needs innovative arrangements between local and traditional leaders to address these concerns.

8.3.7 *Developing local strategies*

South Africa has appropriate tools for developing local strategies, known as integrated development plans (IDPs). Mining legislation requires local governments to dovetail these with the mining companies' social and labour plans (SLPs). SLPs provide some guidance for planning closure, but the guidelines are poorly applied. Furthermore, these plans seldom mention the dependencies created by mining. The dovetailing of IDPs and SLPs can create more dependencies. The outcome is inappropriate closure plans. But there is a more compelling reason why these plans are often inappropriate for planning for closure: they usually portray only one scenario, one that assumes prosperity will continue. The reality of planning in the dark does not feature (Nel & Oranje, 2022). I argue that an appropriate strategic planning approach for mining areas should consider different scenarios. This is important because of the uncertainty of mining and the need to prevent long-term dependencies.

Mining is a temporary activity and brings with it a large degree of uncertainty. The development of scenarios is the first step to understanding uncertainty. It could also be the first step towards getting away from one-dimensional planning systems focusing on growth, prosperity and modernity in mining communities. Does this mean that local governments should ignore the pressures of growth? I do not think so. But they must understand that mining booms and busts are not linear or simple processes, that their nature and scale can vary and that local planning processes should understand the complexities. Planners must consider several scenarios.

Looking at a variety of scenarios can prevent long-term dependencies. It could help identify existing dependencies between mining and local governments. For example, a municipality can become dependent on the property rates and taxes it receives from mining companies and mining households. In a decline scenario, the question is how the local government will recoup costs or minimise expenditure. Scenarios should also contain ways to minimise infrastructure that can become redundant in the case of decline. In the context of housing policy, scenarios could consider the role of rental housing in preventing long-term liabilities for households or local governments.

Mining companies do scenario planning. They have largely incorporated closure planning into their mine life-cycle planning. Within this closure planning, they usually develop a range of scenarios. They adjust their closure plans regularly by considering the mining reserves and market outlooks. They seldom share this information. There is no reason why local governments cannot follow their lead and also weigh the basic information about mining and markets. This would require careful thinking about what different scenarios would bring. It is possible, of course, to get scenario planning wrong. The Free State Goldfields did scenario planning at the end of the 1980s. Within ten years after the decline set in, the situation was worse than their most negative scenario (Marais, 2013b). Scenario planning could help ensure that the likelihood of closure is integrated much earlier into the planning process and that the dependencies that mining creates are identified and considered as part of mine closure planning.

8.3.8 *Developing a local planning framework to help manage mine decline*

One of the main problems with planning decline at the local level is that nobody thinks mine decline is a reality until it occurs. IDPs and SLPs are regional planning instruments for dealing with the planning consequences of mining. This form of collaborative planning has been criticised extensively (e.g. see Van der Watt & Marais, 2021). However, there is no reason to think that collaborative planning is per se inappropriate.

One of the biggest contradictions in the mining industry is the huge amount of effort that companies put into obtaining a social licence to operate compared with how little effort they make to plan adequately for the social consequences of mine closure. Yet, the guidelines for neither the SLPs nor the IDPs deal at any length with concerns about mine closure. Such a planning framework requires a sensitive mapping of the implications of mine closure on three levels: household, economy and city management. Most of these local plans assume continued economic growth. There is an urgent need to develop appropriate planning guidelines for economic decline.

8.4 Further research

There are several areas of further research that should receive high priority. Working through the data and the various arguments, I noted several shortcomings, particularly the shortage of data for understanding mineworkers and their towns or cities. I identified the following areas that demand further research on housing policy and mine closure in South Africa.

8.4.1 Housing, health and productivity: the link

The link between housing, health (including mental health) and productivity requires urgent attention. Much research is available on the poor living conditions of mineworkers. The Marikana massacre in 2012 was sparked by an outpouring of mineworker anger at these conditions. Yet, far less work is available on the economic value for mining companies of ensuring appropriate housing and good living conditions and how this could reduce mine accidents and increase productivity.

8.4.2 Mining and dependencies

As I have emphasised earlier, much of the social investment by mining companies goes into obtaining a social licence to operate. Mining companies have to submit an SLP as part of their mining licence application. Consequently, analysts often refer to the front-end approach in mining. In this front-end approach, most investments take place before mining starts. In this book, I have highlighted the long-term dependencies that mining creates. Many social investment programmes and programmes to ensure a social licence contribute to these dependencies. Examples are clinics run by mining companies. At closure, the mines transfer these clinics to the Department of Health. However, there is no guarantee that the Department of Health is fully prepared to handle the transfer. Research on mine closure in Africa has shown that the continuation of social services originally run by mining companies becomes a problem when mines close.

8.4.3 Mine closure plans

Although I reflect on mine closure plans in this book, this is not the book's main topic. I referred to the role that SLPs play in planning mine closure and how local realities influence this role. We would expect municipalities' IDPs to do the same. Consequently, two research needs exist. First, there is a need to document what the existing SLPs and IDPs actually say about mine closure. This research could record perceptions about closure and how mine closure will affect mining towns. This could be done with key informants in government and mining companies and also with households. There is an urgent need for qualitative research in this area. Second, there is a need to understand the dependencies that mining creates in these mining towns. These studies should investigate physical infrastructure, rates and taxes, social infrastructure, schools and health services.

8.4.4 The need to understand mineworker experiences

I have largely used the household surveys in each of the case study areas for this book, although I also used data from interviews with key informants. However, understanding mineworker families and communities requires much more than household surveys. There is ample room for qualitative and even ethnographic work in mining communities and on mineworkers losing their jobs in the industry.

8.4.5 The need for longitudinal work

In this book, I used data from three household surveys in Emalahleni, Rustenburg and Tsantsabane. Household surveys have their limitations. The general absence of longitudinal data on mining households is a major concern in South Africa. The well-known National Income and Distribution Survey (NIDS) dataset records only a very small number of mining households. The three datasets underlying this book are valuable resources for building on and developing a longitudinal understanding of mineworker households and the change over time. There will be value in tracing changes through panel surveys.

References

Atienza, M., Arias-Loyola, M. & Phelps, N., 2021. Gateways or backdoors to development? Filtering mechanisms and territorial embeddedness in the Chilean copper GPN's urban system. *Growth and Change*, 52, pp. 88–110.

Bainton, N. & Holcombe, S., 2018. A critical review of the social aspects of mine closure. *Resources Policy*, 59, pp. 368–478.

Boschma, M., 2017. Relatedness as driver of regional diversification: a research agenda. *Regional Studies*, 51(3), pp. 351–364.

Brown, R., Dorins, F. & Krannich, R., 2005. The boom-bust-recovery cycle: dynamics of change in community satisfaction and social integration in Delta, Utah. *Rural Sociology*, 70(1), pp. 28–49.

Cloete, J., Venter, A. & Marais, L., 2009. Breaking New Ground, social housing and mineworker housing: the missing link. *Town and Regional Planning*, 54, pp. 27–36.

Digby, C., 2012. Mine closure through the 21st century looking glass. In: A. Fourie, ed. *Proceedings of the 7th International Conference on Mine Closure*. Perth: Australian Centre for Geomechanics, pp. 33–38.

Esterhuyse, S. & Buscke, F., 2022. Coal and water: exploring one precious natural resource at the expense of another. In: L. Marais, P. Burger, M. Campbell, S. Denoon-Stevens, & D. Van Rooyen, eds. *Coal and Energy in South Africa: Considering a Just Transition*. Edinburgh: Edinburgh University Press, pp. 73–87.

Fitjar, R. & Timmermans, B., 2019. Relatedness and the resource curse: is there a liability of relatedness? *Economic Geography*, 95(3), pp. 231–255.

Huchzermeyer, M., 2021. A critical Lefebvrian perspective on planning in relation to informal settlements in South Africa. *Town and Regional Planning*, 79, pp. 44–54.
Haslam McKenzie, F., 2020. Long-distance commuting: a tool to mitigate the impacts of the resources industries boom and bust cycle? *Land Use Policy*, 91, p. 103932.
ICMM (International Council on Mining and Metals), 2019. *Integrated Mine Closure: Good Practice Guide*, 2nd edn. London: ICMM.
Koivurova, T. et al., 2015. 'Social license to operate': a relevant term in Northern European mining? *Polar Geography*, 38(3), pp. 194–227.
Marais, L., 2013a. The impact of mine downscaling on the Free State Goldfields. *Urban Forum*, 24(4), pp. 503–521.
Marais, L., 2013b. Resources policy and mine closure in South Africa: the case of the Free State Goldfields. *Resources Policy*, 38, pp. 363–372.
Marais, L., Burger, P., Campbell, M., Denoon-Stevens, S. & Van Rooyen, D., 2022. *Coal and Energy in South Africa. Considering a Just Transition*. Edinburgh: Edinburgh University Press.
Marais, L., Campbell, M., Denoon-Stevens, S. & Van Rooyen, D., 2021a. *Mining and Community in the South African Platinum Belt: A Decade after Marikana*. New York: Nova Publishers.
Marais, L., Haslam McKenzie, F., Deacon, L., Nel, E., Van Rooyen, D. & Cloete, J., 2018. The changing nature of mining towns: reflections from Australia, Canada and South Africa. *Land Use Policy*, 76, pp. 779–788.
Marais, L. & Nel, E., 2016. The dangers of growing on gold: lessons from the history of the Free State Goldfields, South Africa. *Local Economy*, 31(1–2), pp. 282–298.
Marais, L., Nel, V., Rani, K., Van Rooyen, D., Sesele, K., Van der Watt, P. & Du Plessis, L., 2021b. Economic transitions in South Africa's secondary cities: governing mine closures. *Politics and Governance*, 9(3), pp. 381–392.
Marais, L., Owen, J., Kotze, T., Nel, P., Cloete, J. & Lenka, M., 2021c. Determinants of place attachment among mineworkers: evidence from South Africa. *The Extractive Industries and Society*, 8(3), p. 100943.
Marais, L. & Venter, A., 2006. Hating the compound, but … Mineworker housing needs in post-apartheid South Africa. *Africa Insight*, 36(1), pp. 53–62.
Matebesi, S., 2020. *Social Licensing and Mining in South Africa*. London: Routledge.
Matebesi, Z. & Marais, L., 2018. Social licensing and mining in South Africa: reflections from community protests at a mining site. *Resources Policy*, 59, pp. 371–378.
Measham, T. & Zhang, A., 2019. Social licence, gender and mining: moral conviction and perceived economic importance. *Resources Policy*, 61, pp. 363–368.
Nel, V. & Oranje, M., 2022. Planning in the dark. In: L. Marais, P. Burger, M. Campbell, S. Denoon-Stevens, & D. Van Rooyen, eds. *Coal and Energy in South Africa. Considering a Just Transition*. Edinburgh: Edinburgh University Press, pp. 193–204.
Ntema, J., Marais, L., Cloete, J. & Lenka, M., 2017. Social disruption, mine closure and housing policy: evidence from the Free State Goldfields, South Africa. *Natural Resources Forum*, 41(1), pp. 31–40.
O'Connor, C. & Ruddell, P., 2021. After the downturn: perceptions of crime and policing. *The Canadian Geographer*. https://doi.org/10.1111/cag.12671.
Owen, J. & Kemp, D., 2013. Social licence and mining: a critical perspective. *Resources Policy*, 38(1), pp. 29–35.

Pelders, J. & Nelson, G., 2018. Living conditions of mine workers from eight South African mines. *Development Southern Africa*, 36(3), pp. 265–282.

Ruddell, R., Jayasundara, D., Mayzer, R. & Heitkamp, T., 2014. Drilling down: an examination of the boom-crime relationship in resource-based boom communities. *Western Criminology Review*, 15(1), pp. 3–17.

Swilling, S., Musango, J. & Wakeford, J., 2016. Developmental states and sustainability transition: prospects of a just transition in South Africa. *Journal of Environmental Policy and Planning*, 18(5), pp. 650–672.

Van der Watt, P. & Marais, L., 2019. Normalising mining company towns in Emalahleni, South Africa, *The Extractive Industries and Society*, 6(4), pp. 1205–1214.

Van der Watt, P. & Marais, L., 2021. Implementing social and labour plans in South Africa: reflections on collaborative planning in the mining industry. *Resources Policy*, 71, p. 101984.

WHO (World Health Organisation), 2012. A conceptual framework for action on the social determinants of health. Social Determinants of Health Discussion Paper 2. Geneva: WHO.

Wixe, S. & Andersson, M., 2017. Which types of relatedness matter in regional growth? Industry, occupation and education. *Regional Studies*, 51(4), pp. 523–536.

Index

Note: **Bold** page numbers refer to tables; *italic* page numbers refer to figures.

For Product Safety Concerns and Information please contact our EU
representative GPSR@taylorandfrancis.com
Taylor & Francis Verlag GmbH, Kaufingerstraße 24, 80331 München, Germany

www.ingramcontent.com/pod-product-compliance
Ingram Content Group UK Ltd.
Pitfield, Milton Keynes, MK11 3LW, UK
UKHW021307260926
475702UK00048B/825
* 9 7 8 1 0 3 2 2 0 0 5 5 2 *